W9-AXU-290

HOW TO BUILD
Animal Housing

HOW TO BUILD
Animal Housing

60 Plans for Coops, Hutches, Barns, Sheds, Pens, Nest Boxes, Feeders, Stanchions, and Much More

CAROL EKARIUS

Storey Publishing

The mission of Storey Publishing is to serve our customers by publishing practical information that encourages personal independence in harmony with the environment.

Edited by Deborah Balmuth and Marie A. Salter

Copyedited by Michael Ashby

Art direction by Lisa Clark

Cover design by Kent Lew

Cover photography by © Grant Faint/The Image Bank/Getty (top); © Larry Lefever/Grant Heilman Photography, Inc. (bottom left); © Kent Lew (center right and spine); © Photodisc/Getty (center left); © Sabine Stuewer (center); © Ben Weddle/Midwestock (bottom right).

Finished building illustrations © by Elayne Sears; CAD drawings by Shaun Batho, TSB Consulting; other plans and technical drawings by Brigita Fuhrmann

Text design by Margaret Birnbaum

Text production by Cindy McFarland

Indexed by Jan Williams

© 2004 by Carol Ekarius

All rights reserved. No part of this book may be reproduced without written permission from the publisher, except by a reviewer, who may quote brief passages or reproduce illustrations in a review with appropriate credits; nor may any part of this book be reproduced, stored in a retrieval system, or transmitted in any form or by any means — electronic, mechanical, photocopying, recording, or other — without written permission from the publisher.

The information in this book is true and complete to the best of our knowledge. All recommendations are made without guarantee on the part of the author or Storey Publishing. The author and publisher disclaim any liability in connection with the use of this information. For additional information, please contact Storey Publishing, 210 MASS MoCA Way, North Adams, MA 01247.

Storey books are available for special premium and promotional uses and for customized editions. For further information, please call 1-800-793-9396.

Printed in the United States by McNaughton & Gunn, Inc.
20 19 18 17 16 15 14 13 12 11

Library of Congress Cataloging-in-Publication Data

Ekarius, Carol.
  How to build animal housing / Carol Ekarius.
    p. cm.
  Includes index.
  ISBN 978-1-58017-527-2 (pbk. : alk. paper)
    1. Livestock—Housing—Design and construction. I. Title.
TH4930.E33 2004
636.08'31—dc22

FOR ALFRED R. EKARIUS

(July 3, 1929–September 14, 2002)

He could build anything;

he could fix anything;

he taught me to try

and encouraged me to live the life I wanted.

CONTENTS

PREFACE

This book is not meant for the industrial agriculture folks who see keeping a hundred thousand chickens under one roof as a great achievement. My goal with this book is to help people who, like Ken and me, fall into the need-animals-in-our-lives group, and particularly those folks who are just starting down the path to animal nirvana.

We want to offer a home for our four-legged and winged friends that's both inviting and safe for them, but that also meets our needs for personal safety, good value, and convenience. We've raised all kinds of animals, encouraged a wildlife-friendly environment on our farm and ranch, and done all kinds of construction projects. Over a couple of decades, we've learned a great deal about caring for animals and about providing them with the right home.

To paraphrase gardening guru Eliot Coleman, we strongly believe in the values and rewards of the small farm, and we wish to encourage them. Accordingly, most of the emphasis is on housing and shelter for small-scale livestock endeavors.

As with any book, there are many people to whom I owe thanks: First and foremost, I owe thanks to Ken. Not only is he my best friend and true love, he is also our in-house general contractor (I'm the gopher, hold-er-up-er, and girl Friday for him on construction projects — though I do take credit as the resident electrician).

Thanks also to my friends, the crew of the Coalition for the Upper South Platte. They have carried an extra burden — and taken on additional responsibilities — so I would be free to work on this book.

The staff at Storey is supportive and great to work with. Writers hear horror stories of publishers being tough to work with, but the Storey crew has always been great to me. For this project, particular thanks go to editors Deborah Balmuth and Marie Salter, and to artists Elayne Sears, Shaun Batho, and Brigita Fuhrmann, who took my scribbles and notes and made sense of them, creating the great illustrations and plans you see here. I also thank Dr. David Kammel, a state Extension and research specialist for livestock housing at the University of Wisconsin, and Dan Herman, a general contractor in Colorado Springs, Colorado, for reviewing the text for me.

We all owe our government a debt of thanks for supporting institutions like the land-grant colleges, the Cooperative Extension Service, the U.S. Forest Service Forest Products Laboratory, and other research organizations. Little of what I have written in this book would have been possible without decades of research and publication by these organizations; much of the information you read here, and most of the plans you see here, are directly attributable to the efforts of their scientists and engineers.

I invite you to visit my Web site often for information on my books and other writing: www.carolekarius.com. From the main page, click E-MAIL ME to drop me a line, or click Q&A to post questions that I will answer on the site.

I. PREREQUISITES

We've all heard the saying "Fools rush in where angels fear to tread." If you're like me, you probably get excited about starting projects and want to jump right in, but only a fool would rush to start an animal housing project; haste of any kind costs money and time, and tends to yield a less-than-satisfactory result. So take a lesson from the angels: don't rush. Take plenty of time to think and plan before starting your project.

The chapters in this section are meant to assist you during the planning phase. Studying them will pay long-term dividends for you and your animals, because you will be better able to address important animal health concerns. Adequate ventilation, appropriate flooring, and selecting building materials that are not toxic to animals are all critical; take none of these issues lightly.

Chapter 1, Shelter, talks about the housing needs of animals and addresses health and safety issues. Chapter 2, Planning, gets you thinking about budget, legal issues, and how to work with a contractor. Chapter 3, Structure, Design & Function, provides background information on topics such as ventilation (one of the most important considerations in animal housing according to veterinarians and agricultural engineers), heating, insulation, and bedding. Animal housing can be a major investment, so chapters 2 and 3 also help you address design issues that impact costs and the long-term value of your investment.

Looking ahead, part II, Plans, provides sample plans to inspire the designer in you. Part III, Construction, provides specific construction information for some of the topics introduced here. Be sure to read this material; it will be useful whether you plan to do the work yourself or work with a contractor.

1. SHELTER

The north wind doth blow, and we shall have snow,
And what will the robin do then, poor thing?
He'll sit in a barn, to keep himself warm
And hide his head under his wing.

—TRADITIONAL NURSERY RHYME

Some people are simply driven to have animals in their lives. For those of you who fall into this category, know this: you are in good company. About 85 percent of your friends, neighbors, and relatives fall squarely into the same camp.

For some people, time and space limit their direct contact with critters, but they foster a relationship with wildlife by feeding birds, talking to squirrels, and simply enjoying the natural life in their backyard. For others, family companions like dogs, cats, and birds, or something a little more exotic — ferrets, potbellied pigs, or iguanas, for instance — enrich their world. Then there's the group that Ken

and I fall into: the group that keeps farmyard animals.

The relationship between people and animals began back when people still considered a skin-and-pole structure to be the lap of luxury and their animals sought shelter anywhere they could. In good times, the animals found relief from nature's fury on the lee side of hills, or in forests or arroyos. In bad times, they found no relief from rain or snow, sun or flies.

Now, by and large, we find shelter in airtight, climate-controlled boxes. We are comfortable in shorts and T-shirts during the cold of winter or in our flannel shirts in the heat of summer. Do our animals need the same climate-controlled environment that we call home? No, they don't.

So, before you get ready to build your dream barn, ask yourself if you actually need a building. Seriously. The cost of building a barn or stable is very high, and you may not need one at all. As our ancestors knew, livestock don't necessarily need a permanent building. If your animals will be having their babies on pasture during the spring flush (the early part of the grass-growing season), and if they have shelter from wind and sun, they can get along pretty well without buildings, or they can use portable or temporary structures. Windbreaks and shade structures are easier to build and cost far less than a barn. We have raised all kinds of animals for decades and have done without a barn more often than not.

If, on the other hand, the babies will be born in winter, you will need some animal housing, and small animals (chickens, turkeys, ducks, rabbits) must have some kind of building to keep them safe, especially at night, from marauding neighborhood dogs or scavenging coyotes, but it can be a small and inexpensive shed.

Admittedly, buildings often make *our lives* easier. Taking care of a sick animal in January is much less onerous in a barn than out on pasture during a blizzard, and having an organized building with storage for feed, tack, tools, and equipment, along with a couple of pens for animals that need special attention, is wonderful. But not having a barn doesn't mean you can't successfully raise some happy animals.

ANIMAL WELFARE

As agriculture has moved to an industrial, or factory-farm, system, with vertical integration (one company controlling the animal from birth to the consumer's plate) and confinement as its cornerstones, animals have come to be treated like machines. Two hundred thousand hens may occupy a single building at an egg farm, each allotted less space than the page you are reading, and hogs are typically raised in a building the size of a football field, with thousands of animals crammed inside.

The impacts of factory farms include cruelty to the animals, environmental problems that reach far from their source, and social problems for rural communities that lose their

Factory Farms and the Environment

In 1950, there were 2.1 million farmers selling hogs in the United States. Fifty years later, in 2000, there were fewer than 99,000. Yet, while the number of hog farmers has decreased, the number of hogs produced has remained roughly the same. The difference is that the average number of hogs per farm has increased dramatically, from 31 per farm to 1100 per farm.

Today, there are 105 farms producing more than 50,000 hogs each, accounting for 40 percent of the U.S. hog inventory. And four corporations account for 20 percent of total U.S. hog production.

A single hog, raised industrially, produces 1.9 tons of manure per year. A typical 5000-hog factory produces 9500 tons of manure per year, or nearly 26 tons per day — about the same amount generated by a small town. But a town would have a sewage-treatment plant, and traditional, small-scale hog farms integrate manure into the farming system by composting it and using it to fertilize fields, without nutrient runoff. Hog factories, on the other hand, store these massive amounts of waste in open lagoons or, increasingly, in concrete cisterns under the hog factory barns.

When this manure runs off the land or leaks into waterways, fish kills can occur. Local well water can become contaminated, and massive swarms of flies can emerge. What's more, odors from lagoons can make being outside unbearable for people who live near a hog factory.

Iowa, which produces 15 million hogs a year (more than any other state), constitutes only 5 percent of the Mississippi River watershed but contributes 25 percent of the nutrient pollution to the waterway. Nutrient pollution has created a 7000-square-mile (and growing) dead zone in the Gulf of Mexico [a dead zone is an area that doesn't have enough oxygen in the water to support the native sea life].

The cumulative impact of many smaller spills can be equally harmful. In 1999 alone, one of the largest pig farms in the United States was responsible for 25 liquefied manure spills and discharges in Missouri. More than 224,000 gallons of manure and wastewater were discharged during these spills. (The Humane Society of the United States, "Communities and the Environment." http://hsus.org/ace/15084 [accessed Feb. 10, 2004]. Reprinted with permission; © 2004.)

The Five Freedoms

Animal welfare advocates support the five freedoms for domestic animals:

1. Freedom from malnutrition

2. Freedom from discomfort

3. Freedom from disease

4. Freedom from fear or distress

5. Freedom to express normal behavior

base of small, independent farmers. To deal with health problems associated with large confinement operations, factory farms routinely use antibiotics, exposing consumers to minute doses of these drugs, which is believed to contribute to the reduced efficacy of many antibiotics in treating human illness. These issues have helped spur strong support for animal welfare among consumers, and small farmers have an advantage in that they can market humane animal treatment to great advantage.

Stress and Fear

When animals are stressed or scared, they demonstrate physiological and behavioral changes, ranging from increased respiration to aberrant behaviors like cage biting or fighting. Healthy, happy animals, on the other hand, tend to exhibit few of these changes. Dr. Temple Grandin, a renowned animal scientist at Colorado State University, says that animal cruelty falls into two categories:

CATEGORY 1. ABUSE AND NEGLECT

These are abuses that good livestock producers would not tolerate. They are animal cruelty abuses such as dragging downed crippled cattle, rough handling, throwing baby dairy calves, beating an animal, starving an animal, failing to provide shelter, or shackling and hoisting an animal prior to ritual slaughter. Almost all problems that occur during handling, transport, and slaughter of livestock are category 1 abuses. I estimate that over 75 percent of all livestock producers, transporters, and slaughter plants do a good job of preventing these abuses. However, 10 percent allow category 1 abuses to occur frequently and another 10 percent occasionally have problems with animal abuse. This is an area where the industry needs to clean up its house and take action against the bad operators.

CATEGORY 2. BOREDOM AND RESTRICTIVE ENVIRONMENTS

Whereas the animal welfare issues in category 1 concern obvious animal abuses and cruelty, the issues in category 2 do not involve pain. Category 2 welfare issues are animal boredom and abnormal behaviors, which may occur in barren environments that do not provide adequate stimulation. Examples would be gestation stalls for sows, veal calf housing in individual stalls, and chickens in cages. (Temple Grandin, "Dr. Grandin Speaks Out on Animal Welfare Issues." http://www.grandin.com/welfare/welfare.issues.html. [accessed Feb. 10, 2004].)

Thankfully, I have rarely seen the category 1 abuse Dr. Grandin describes, but I have seen many examples of category 2 abuses on farms and in backyards. Although perhaps less repugnant than category 1 cruelty, boredom and restrictive environments are cruel nonetheless. Simple steps — such as providing adequate space and an opportunity to go outside; light, airy quarters; and opportunities for socialization with other animals — can go a long way in reducing boredom

Understanding Animals

Understanding how animals respond to their world through their senses will help you understand what causes stress and how it can be minimized through effective housing design and management. Livestock species are prey animals and, as such, they have developed senses that help them find food and shelter, navigate around their range, and, most important, avoid predators.

Most farm animals have a wider field of vision than humans, which allows them to see predators moving in on them from almost any direction. Their eyes are large relative to their head size and are located on the sides of the head, allowing them to see different views with each eye. (The one exception is the pig, which, as an omnivore, has eyes very similar to our own.) In spite of the wide field of vision, most animals have a blind spot directly behind them, so use caution when approaching animals from the rear.

Visual acuity is the most relied-on sense for prey animals, followed closely by hearing. New items in a pen or barn — say a jacket hanging over a rail or a shovel lying in the middle of the walkway — can cause a fear response. Loud, sudden noises and high-pitched sounds also cause a fear response in most animals.

Like people, animals have a natural curiosity about the world around them, and they tend to use their senses of smell and touch to investigate. Through touch and smell, they identify their offspring and herd mates and explore new things, like a tractor parked in the pasture and a new type of feed.

Most animals adapt quickly to new things if given the chance to investigate them, so be sure to allow for investigation. Whenever you introduce them to something or someplace new, let them take their time checking it out. An animal with a calm disposition that has never suffered from extreme human-caused fear may seem to immediately accept the novel in its world; animals that are high-strung or that have had bad experiences may require some time to settle down and accept what you have introduced to them. Be patient to avoid risk of injury to you or your animals.

SAFETY

The first consideration in animal housing is safety for you and your animals. Tripping and falling are the most common types of accidents. Floors should have a nonslip surface. For example, if animals will be expected to walk on concrete floors, the concrete must be grooved for traction (½-inch-deep grooves work well), coated with a nonslip coating, or covered with rubber mats or deep bedding.

Space should be provided so that tools and equipment can be stored out of walkways, and storage areas (especially those where feed, medicines, or chemicals like cleaning supplies are stored) should be constructed so animals cannot access them. Walkways should be at least 10 feet wide for large animals, such as horses and cattle, and 7 feet wide for smaller ones, like sheep, goats, pigs, and llamas.

It is *critical* that electrical installations be inaccessible to animals, and systems must be designed for the humid and dusty conditions typical in barns and stables. Lighting should be adequate for you and your animals to be able to move with clear vision. Light fixtures should be installed so that lighting is even and diffuse; the fixtures should have protective, break-resistant covers that prevent dust and moisture from corroding wires and causing shorts. The electrical system should be well grounded and have adequate load capabilities for the tools and equipment that will be used in it. Use ground fault circuit interrupter (GFCI) plug receptacles. Appliances, such as radios and coffeepots, should be unplugged when not in use.

Fire prevention is an important consideration in animal housing. Barns are often full of highly combustible materials, like hay and straw. A carelessly discarded cigarette or match, an electrical short, or a hot engine can set off a conflagration in no time. All barns should have fire extinguishers placed near each door. Consider installing a heat-and-smoke sensor with an alarm mounted outside the barn and one in the house. This new type of sensor picks up a sudden increase in temperature, even if smoke is not yet detectable. Barns can also be designed to include sprinkler systems.

All construction should be done so that there are no protruding nails or exposed metal edges. Gates should have secure latches, yet they should be easy for you to open with one hand or when wearing heavy gloves. Paints and wood preservatives need to be carefully selected, as many of these products can give off toxic fumes or may be toxic if ingested; bored animals often chew on exposed wood.

OTHER CONSIDERATIONS

There are a number of other points to consider in designing animal housing that will contribute to the general health and happiness of your animals. I firmly believe all animals should have access to pasture or outdoor pens, which is the easiest way to avoid the boredom of category 2 abuse, but those that will spend a significant amount of time in a building need sufficient space, clean air and good ventilation, clean water, comfortable

MINIMUM HOUSING DIMENSIONS

PENS AND STALLS

Animal	Maternity pen (sq. ft.)	Box/stall-type pen (sq. ft.)	Ceiling height (ft.)	Minimum space per animal at shared feeders
Cow, dairy	144	50	10	3 ft.
Cow, beef	128	45	8	2 ft.
Heifer	NA	35	7	1 ft.
Bull	NA	100	10	4 ft.
Finishing cattle, 800–1200 lbs.	NA	35	7	2 ft.
Calves, 400–800 lbs.	NA	25	7	1.5 ft.
Calves, less than 400 lbs.	NA	15	6	0.75 ft.
Mare	144	100	10	3 ft.
Stallion	NA	196	10	6 ft.
Foal	NA	100	7	2 ft.
Ewe or doe	25	16	6	1.5 ft.
Ram or buck	NA	30	6	1.5 ft
Lamb or kid, preweaned	NA	2	6	2 in.
Lamb or kid, weaned	NA	10	6	1 ft.
Sow	64	48	7	2 ft.
Gilt	NA	40	7	1 ft.
Boar	NA	60	7	3 ft.
Finishing pigs, 150–220 lbs.	NA	16	6	1.5 ft.

NEST BOXES AND LIVING SPACE

Animal	Nest boxes (sq. ft.)	Floor space per bird (sq. ft.)	Perch space per bird (ft.)
Hens, layers	1.5	1.5	0.75
Broilers	NA	1.5	0.75
Bantam chickens	1	1	0.5
Turkey, heavy toms	NA	5	NA
Turkey, toms	NA	4	NA
Turkey, hens	3	3	NA
Turkey, broilers	NA	2	NA
Pheasants	3	5	NA
Ducks	2	3	NA
Geese	3	6	NA

NA = not applicable.

bedding or flooring, and freedom from excessive noise. Handling facilities, to ease moving animals into and out of buildings and for quarantining new animals, are an important component of an overall building plan.

AIR QUALITY AND HEALTH

Nothing can kill an animal quicker than being locked up in a poorly ventilated building. Animals in a barn give off moisture and gases such as carbon dioxide and methane as part of their respiratory and digestive functions. Manure and urine add ammonia, hydrogen sulfide, and carbon monoxide to the air. Dust from bedding and feed adds particulates to the air that aggravate respiratory passages and may host allergen-producing molds. And pathogenic bacteria and viruses can survive in moisture-laden air for long periods.

Ventilation systems can be either natural or mechanical. They should be designed to remove excess moisture and contaminants, prevent drafts, and maintain comfortable temperatures for the age and class of animal spending time in the barn. They should also have both inlets for bringing in fresh air and outlets for removing stale air. The best ventilation occurs when there is a good balance of these two types of vents. (See page 32 for more on ventilation.)

COMMON GASES & THEIR HEALTH EFFECTS

Gas	Odor	Effects
Ammonia (NH_3)	Sharp, pungent	Irritates mucous membranes (eyes, throat, nasal passages), coughing, frothing, difficult breathing, fluid accumulation in lungs resulting in pneumonia-type illness, blindness due to corneal burning
Carbon dioxide (CO_2)	None	Drowsiness, irritability, difficulty breathing, chest pain
Carbon monoxide	None	Dizziness, headache, flulike symptoms, nausea, irritability, drowsiness
Hydrogen sulfide (H_2S)	Rotten eggs	Irritates mucous membranes (eyes, throat, nasal passages), headache, nausea, insomnia, loss of appetite, fluid accumulation in lungs
Methane (CH_4)	None	Headache, difficult breathing, irregular heartbeat, staggering gait, convulsions, coma

Note: These symptoms result from short-term exposure to high concentrations or from long-term exposure to lower concentrations. These gases can affect all classes of livestock, and people, at varying exposures and concentrations. All are potentially fatal in high enough concentrations or can cause serious long-term health problems with regular exposure at lower concentrations.

2. PLANNING

Although pre-project planning isn't a great deal of fun, if you are investing in permanent structures, it's important to spend some time on planning. In fact, it is probably the best investment you will make on a major construction project. It saves money by avoiding costly mistakes and also time, that precious commodity we never seem to have enough of. Planning is a process, but it's easy if taken in a step-by-step fashion.

MONEY MATTERS

Because Ken and I have been self-employed for the last couple of decades and are outside the 9-to-5-job-with-a-good-paycheck world, money is always on our minds. We live very well and happily in a place we love, but all capital expenditures require serious thought and debate. As I said in the first chapter, you may not even need animal housing, or you may be able to get by with something small, temporary, or portable, but if you decide you want a barn or stable, remember that the costs don't end when the project is done.

As you consider the cost of your project, also consider the following long-term expenses:

Interest on borrowed funds can be with you for years. Even with low interest rates, a twenty-year loan can easily double the cost of your investment, so try to pay off debts as quickly as possible — and shop around for the best rates. Tying a loan to your mortgage will give you the best rate but will also extend the term. Most mortgages have early-payment options, and by paying a little extra each month, you can significantly reduce the total cost over the life of the loan.

Property taxes, which are assessed based on the property's actual value, help pay for local government services and schools. Permanent improvements add to your property's value, so you will pay additional taxes on them; temporary or portable units are usually exempt from property-tax assessment, but local governments may have a "business personal property" or real property tax in place that could apply to these structures if they are accessories of an operating commercial enterprise.

PREPARING AN ESTIMATE

For most loans, you will need to prepare a cost estimate before you apply, but even if you are funding the project from cash on hand or a credit card, producing an estimate is a valuable exercise. The table that follows may help you in preparing your estimate.

Estimate Essentials

Item	Units (ft., sq. ft., each, etc.)	Cost per unit	Total estimated cost	Actual cost
Loan costs (origination fees, points, etc., for second mortgages, refinancing)				
Liability insurance				
Building permits				
Temporary utilities (check with your electric company)				
Plans and specifications				
Excavation, clearing, grading				
Footings				
Foundation				
Concrete for slab				
Framing lumber				
Windows				
Doors				
Roofing				
Siding and trim				
Heating, venting, insulation				
Utilities (electrical supply, water, and sewer/septic, if appropriate)				
Plumbing				
Electrical (break out by class: e.g., outdoor lights, junction boxes, and wire)				
Painting				
Equipment rental				
Waste disposal				
Landscaping				

These two graphs illustrate the impact of various loans on monthly payments and on total payments over the life of a loan. We are assuming in this example that someone is borrowing $40,000 for a barn project, and we compare three different interest rates (5%, 7%, 9%) and four different payment periods (5, 10, 15, 20 years). As the term of the loan (in years) increases, the payment per month drops dramatically, but the total payment over the life of the loan increases just as dramatically.

If you have access to a computer with Microsoft Excel, you can make comparisons for different rates and terms you are considering by doing the following: In column A, enter the amount that will be borrowed; in column B, enter the interest rate divided by the annual payments (for example, a 5 percent loan with monthly payments is entered =5%/12; a 5 percent rate with quarterly payments is entered by keying =5%/4); in column C, enter the term in payment periods (for example, 5 years of monthly payments =5*12; 10 years of quarterly payments =10*4). Now in column D, enter the following formula: =PMT(B1, C1, A1). This will be the amount of your payment (not including taxes, reserve payments, or other fees that may be associated with a loan). To get total payments for the life of the loan, set up an equation in column E that will multiply the result in column D by the number of payments: =D1*C1. (To apply formulas to successive rows, click on the cell, place the cursor over the black square in the lower right corner of the cell, then click on it and drag it down.)

You can do the same thing with Lotus 1,2,3, but instead of keying in the equals sign, key in the *at* symbol (@) and enter the order of functions in the parentheses as A1, B1, C1. Thus, you would enter @PMT(A1, B1, C1).

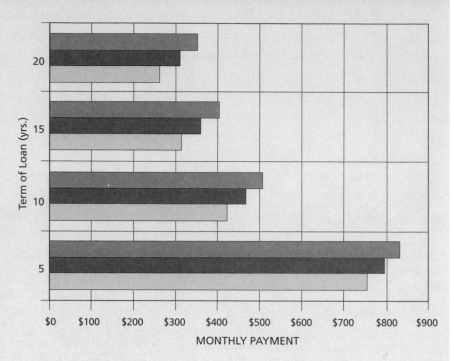

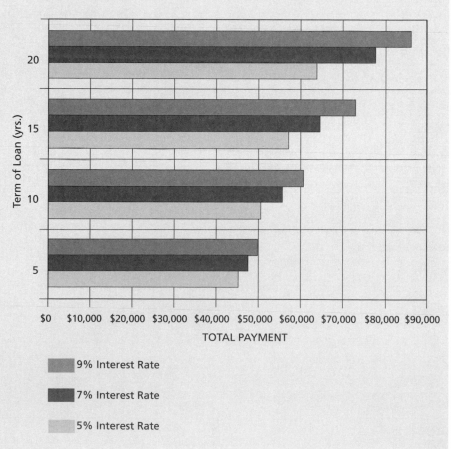

9% Interest Rate

7% Interest Rate

5% Interest Rate

Maintenance is an expected and ongoing cost. Early in a new building's life, this cost may be minimal, but rest assured it will crop up quickly and grow over time. With their chewing, kicking, and bashing, animals can be hard on buildings. Moisture in animal housing causes deterioration of wood and rusting of metal surfaces.

Utilities can eat a healthy chunk of your income, particularly if you will be using fans for ventilation in a confinement building. A dairy farmer friend of ours in Minnesota knocked off about $1000 per month from his summer utility bills when he went from a confinement to a grass-based operation: his cows were out on pasture most of the day and so he wasn't running large ventilation fans all day long.

Insurance costs come with permanent structures, but by speaking to your agent ahead of time, you may discover ways to reduce these costs. For example, most insurance companies offer discounts for buildings that include smoke and heat detectors.

Depreciation is the "costing out" of an asset over time and is largely a tax-related issue (a building constructed as part of a commercial agricultural enterprise can provide a depreciation tax deduction), but savvy businesspeople know that depreciation also has a planning component to it. As an asset depreciates off the books, its replacement or renovation becomes a factor, so businesses often "bank" depreciation as they take it, thus allowing the replacement or renovation to take place with cash when the need arises.

Opportunity cost is the amount you could earn with the money otherwise tied up in a building were it invested elsewhere. Treasury bonds are a good benchmark for safe earnings; they provide a secure, minimal return on investment. Calculate your opportunity cost by figuring how much interest you would earn if you invested the money you are considering spending on a building.

SITE SELECTION

Begin the planning process by collecting maps and aerial photos of your site. They will supply you with a wealth of information on soils, drainages, and other natural features. Next, draw a working site map to scale: show the property lines, existing structures, roads and driveways, public utilities, prevailing wind directions for both winter and summer, and such natural features as ponds, wetlands, hard-rock outcroppings, and slopes.

Wetlands larger than one acre are protected by federal laws and require a 404 permit to drain or fill, and you don't want to go there. Even if there are no regulations protecting small wetlands in your area, it is best to avoid them: build in a wetland and you will have a permanent drainage problem and possibly animal health problems, such as foot rot and mastitis, that commonly occur on poorly drained soils.

Large rock outcroppings always increase construction costs, so avoid these areas, too, if possible. Slopes of up to 8 percent are fairly easy to build on; although you may incur additional expenses excavating and preparing a site on a slope, building into or downwind of a

> **TIP**
>
> Find topographical maps and aerial photos of your site at TerraServer, a service operated by Microsoft in partnership with the U.S. Geological Survey. (See resources for Web address.)

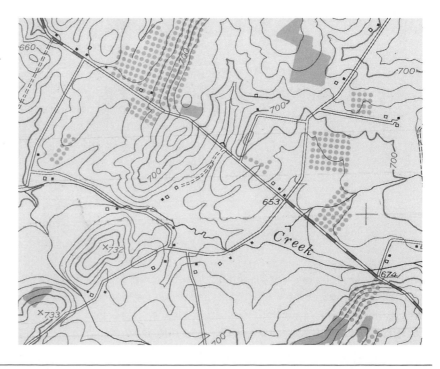

Topographical maps indicate road classifications, bodies of water, and land elevation. (Reprinted from Smithsburg Quadrangle, U.S. Geological Survey, 1994.)

Sources of Maps and Aerial Photos

When beginning your planning process, acquire as many informational maps of your site as you can. Soil and topographic maps are especially helpful. The following sources might be able to supply you with pertinent maps for free or at a low cost.

- Local governments (counties, towns, fire districts) have maps that show roads, section lines, and property lines.
- Local conservation district offices, Natural Resources Conservation Service (NRCS) offices (a U.S. Department of Agriculture agency), and Cooperative

- Extension Service offices have soil maps and may also have maps showing vegetation types.
- State foresters
- State mineral offices
- U.S. Geological Survey
- U.S. Coast Guard if your property borders navigable waters

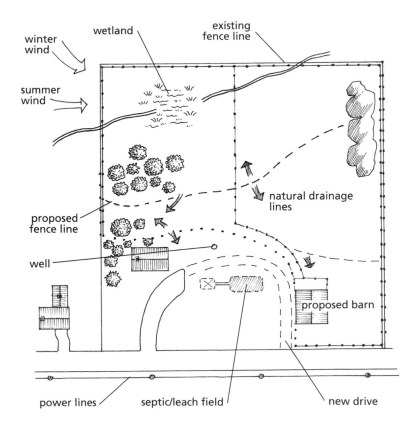

Draw a working site map for planning purposes that shows property lines, existing and proposed structures, roads and driveways, public utilities, prevailing wind directions for both winter and summer, and such natural features as ponds, wetlands, hard-rock outcroppings, and slopes.

slight hill can reduce the effects of winter winds. On slopes of between 8 and 16 percent, you will incur significant additional costs for engineering as well as excavating and shoring. Slopes greater than 16 percent are challenging and therefore very expensive to build on.

In the best of all possible worlds, your home is located at a higher elevation than the housing for your animals. A higher elevation gives you a view so you can see what's going on with your critters from the comfort of the house, and it reduces odors and the risk of water contamination and flooding. Typically, barns are located at least 300 feet from the house and downwind of the prevailing wind direction. Small chicken coops, rabbit sheds, and kennels should be at least 75 feet from the house.

Animal housing shouldn't be built in a wooded area, and, in fact, to protect against fire danger, it should be set back at least 50 feet from trees. This will provide an opportunity for a fire traveling through the crown of a forest to drop to the ground and cool off. If your barn is immediately adjacent to a heavily wooded area and a fire approaches, it *will* consume your barn. If you live in forested country, talk to your state forestry office about *defensible space* recommendations. These include things like the best building materials for fire resistance, driveway construction considerations for fire truck access, and removal of "fuels" like wood piles and brush from around buildings.

Water and Sewerage

Your animals will need a good supply of drinking water, which can come from a municipal water system, a private well, spring, pond, or creek. Unless you live in an area where there is no risk of water lines freezing, bury lines from the well to the barn below normal frost depth, and include a drainage valve on the barn's water supply line so water can be drained from pipes in the building.

Installing yard hydrants is a relatively cheap investment if done when new water lines are being run but an expensive one as an afterthought. Unlike a water pipe with a hose bib attached, yard hydrants are designed to weep water back out of the hydrant between uses, which prevents ice from traveling into the main water line.

Check with your local health department about what the minimum setbacks are for water and sewer components, and consider them minimums. The cost of cleaning up a contaminated water supply will be far greater than the cost incurred to protect the water supply in the first place, so keep all animal yards and buildings (even dog enclosures) away from a wellhead. If you will be drilling a new well for a barn, do so in an elevated area upslope from the barn and any heavily used loafing areas or holding pens. If your county or state has no minimum setbacks from a well, consider 100 feet the absolute minimum.

If you are considering milking cows, goats, or sheep for commercial purposes, your barn will need a plumbed milk room with its own septic system. You'll also need a septic system

WATER REQUIREMENTS FOR OPTIMAL ANIMAL HEALTH

Animal	Water consumption in cold weather (gal./day)	Water consumption in hot weather (gal./day)
Cow-calf pair	10–18	30–40
Dairy cows, lactating	20–40	25–50
Dry cows	10–15	20–30
Calves (per 100 lbs. body weight)	1–4	1.5–8.0
Growing cattle (400–800 lbs.)	4–10	6–18
Finishing cattle (800–1200 lbs.)	6–15	15–25
Bred heifers (800+ lbs.)	8–15	17–25
Bulls	12–16	25–35
Sow and litter	6–8	8–10
Gestating sow, gilt, or boar	3–6	4–8
Nursery pigs	0.5–1.0	1.0–1.5
Growing pigs	1.5–2.5	2.5–4.0
Finishing pigs	3–5	4–6
Ewe and lamb	1.5–2.5	3–4
Dry ewes	0.75–1.25	1.5–2.5
Bred ewes	1.0–1.5	2.5–3.0
Rams	1.5–2.5	2.5–3.5
Horses	6–10	10–15
Chickens 1–4 weeks (per 100 birds)	0.5–1.5	1.0–3.4
Chickens 4–9 weeks (per 100 birds)	1.5–3.2	3.4–6.7
Chickens 9–13 weeks (per 100 birds)	3.2–4.0	4.9–10.0
Laying hens	4.5–8.0	7.5–10.0
Turkeys 1–4 weeks (per 100 birds)	6.5–10.0	7.5–12.0
Turkeys 4–10 weeks (per 100 birds)	9–14	10–18
Turkeys 10–19 weeks (per 100 birds)	12–17	16–28
Rabbits (per 10)	2–4	3–7

if the building will have a restroom. If you are springing for water and septic anyway, consider including a janitor's sink, a washer/dryer, and a shower in the restroom. Being able to shower after a particularly nasty job, or to wash overalls and work clothes, barn rags, saddle blankets, and the like in the barn, is quite a boon.

If your site is near a stream, river, or lake, allow at least a 200-foot setback from the shoreline to your buildings. To minimize pollution and bank erosion, keep this strip in permanent grass. And check into flood insurance.

Drainage

Design roads and buildings on higher ground to prevent flooding, and slope soils slightly away from buildings: muddy barnyards can lead to animal health problems like foot rot and mastitis, and excessive moisture around buildings can lead to failure of their foundations. If structures must be placed in low areas, use "French drains" (perforated pipes buried in a gravel bed that pick up water underground and funnel it out of the area) and gravel or compacted road base as fill to reduce mud. Water can be diverted from buildings by using an earth ridge, terrace, or concrete curbing.

Try to direct runoff to well-grassed, permanent pastures. If you can't move runoff to a grassed area, construct a small, rock-lined drainage pond downslope. Don't allow runoff to approach wells, buildings, or a neighbor's property.

Your local conservation district office, Cooperative Extension agent, or Natural Resources Conservation Service officer can often provide no-cost or low-cost assistance for drainage design, and the NRCS may offer a cost-share program for drainage control if you qualify.

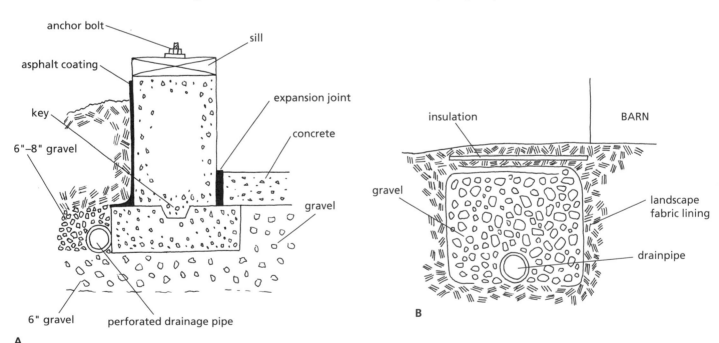

French drains are used to move water away from footings and buildings. Set immediately next to a footer (A), they reduce cracking in concrete. Set a little farther away, they reduce muddy areas in the barnyard. In areas with clay soil, set landscape fabric around the edges of the trench and backfill with clean gravel to prevent the pipe from becoming clogged (B). In areas subject to extreme winter temperatures, place a layer of closed-cell styrene insulation board over the drainage way to help the drain stay open and workable in the spring.

Soil Types

Different soil types have different load-bearing capacity, or ability to disperse the weight of a structure over a given area. Bedrock has the greatest load-bearing capacity, at as much as 40 tons per square foot, and mucky clay soils have the lowest, at about a half ton per square foot. The final design of your building's foundation will depend on the soil's ability to bear the weight of the structure. In areas with unstable soils, you may need a professional engineer to design your structure's foundation. Although bedrock has high load-bearing capacity, it can make installation of underground utilities, like water and sewer lines, a real nightmare.

Soil information is available for many areas of the country from the NRCS office. Call your local NRCS office to find out if it has soil maps for your site.

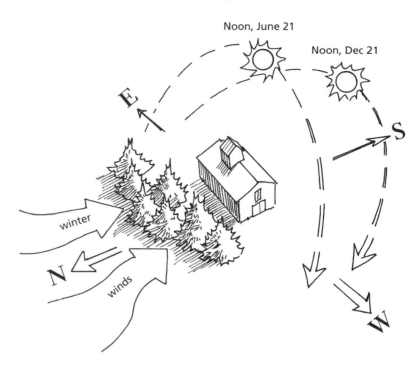

For most of the United States, predominant winter winds are out of the north and northwest. The sun moves higher in the sky during summer.

Sun, Wind, and Snow

In hot climates, choose sites that minimize solar exposure, such as adjacent to trees that block the southern exposure, and place buildings with their shortest side (elevation) to the south. Low rooflines and overhangs on the southern exposure will help to reduce inside temperatures. Place open-faced structures with the opening aligned in the direction of summer breezes for natural cooling.

For cool climates, do the opposite and place buildings so the longest elevation is exposed to the sun. Block winter winds with natural features such as hills and woods or by growing or constructing a windbreak.

Wind can be your friend, helping to naturally ventilate buildings, but it can also be your enemy: think about your downwind neighbors! One of the quickest ways to start a feud is to subject your neighbors to "eau de manure" with any regularity. The more animals you have, the bigger this problem becomes. Wind can also damage buildings and kill animals, particularly very young ani-

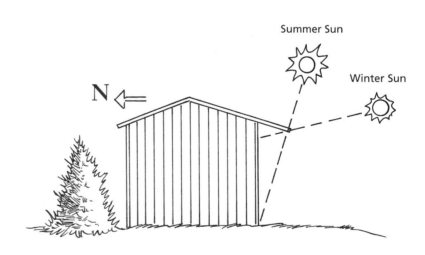

By giving consideration to prevailing wind patterns and seasonal solar paths during planning, you can design and site buildings to take advantage of natural temperature-control strategies, like windbreaks and overhangs.

mals. As you plan, try to site buildings so the prevailing wind moves odors away from you and your neighbors, and consider where windbreaks (natural or man-made) can be incorporated to protect your animals.

Snow removal and storage is not a big issue all over the United States, but if you happen to reside in the snowbelt, determine where you can store plowed snow, and leave enough room around buildings for operating a plow. Windbreaks can reduce snow accumulation in driveways and around buildings.

Access Roads and Utilities

Good roads are expensive to design, build, and maintain, so placing structures where you can readily access them from existing roads is the best plan. When planning, remember that emergency vehicles may need access to your building, so provide at least 75 feet between buildings, with some larger areas suitable for maneuvering fire trucks. Most local fire department officials are more than happy to discuss ways to design your site layout and buildings for fire protection purposes. They want to save your property in an emergency, so utilize their expertise during planning.

At times you may need to accommodate a tractor-trailer in the area around the barn — to deliver bulk feed or transport animals, for example. At other times you may be moving large pieces of farm equipment around buildings. Allow plenty of room for these big rigs. A tractor-trailer requires at least a 55-foot turning radius and 13.5 feet of clearance from overhead obstacles like power lines. Farm equipment may require even more "head room." Running electrical lines underground between buildings removes one overhead concern and improves the appearance of your site.

Animals will need access to pastures, so think about how your animals will move between permanent pastures and buildings. You may need to construct new lanes or holding and training pens, or you may opt to site your building to allow for ready access to existing facilities. Consider whether animals will have direct access to runways that lead to subdivided pastures (also called paddocks). If you are creating runways and lanes with permanent fencing, it's a good idea to make them wide enough to accommodate a tractor or pickup. If you live in an area that gets lots of rain, consider using compacted road base on lanes, on your driveway, and in areas around doors to minimize muddy conditions.

In our neck of the woods, running power poles and wire costs about $12,000 per mile. Although there may be some regional differences in pricing, running power far from existing poles is costly, so consider placing buildings close to existing power lines. And think about power requirements before you begin building. For example, if you are considering installing a milk parlor for a small dairy operation, you will need sufficient power to operate the compressors and vacuum pumps. If you will be keeping your animals confined in the barn, you may need mechanical ventilation with large fans operating full time during the heat of summer.

Your land may have certain deed-based restrictions — such as utility easements — about where you can build, or provisions for accessing underground minerals. Such conditions should be specified on the deed to your property. If there are easements, record them on your map and avoid constructing buildings on them; if someone else controls mineral rights, consult an attorney about your rights.

OTHER CONSIDERATIONS

Once you have narrowed the decision on where your building will be constructed, you have to make some other decisions.

- What type of structure will you build, a traditional barn, for example, or an alternative design?
- Will you purchase a kit or build from scratch; do it yourself or hire a contractor?
- What type of features do you want to incorporate? Do you want to include an office, a bathroom, visitors' or helpers' quarters, feed

storage, or space to park equipment? Special features are relatively easy and inexpensive to include from the start of a project but tough to add later.

- Will you be building a loafing shed (a three-sided structure with free access from a pasture) or a barn? If you are building a barn, will it house animals full time (confinement) or occasionally?
- Are there architectural design features you want to incorporate — like cupolas or porches?
- Do you need a building permit and/or a zoning review?

Reduce, Reuse, Recycle

We are committed to recycling, and over the years have recycled several very old outbuildings as well as an 1860s miner's cabin for residential use. There are some great advantages if you have a building that is structurally sound — or still reasonably capable of being restored to structural soundness — but reusing old buildings takes a real commitment. The cost of rehabilitating an old barn can run from about one-third to more than twice the cost of new construction. It all depends on the barn's structural condition and your level of dedication to the project.

Old barns are an endangered species, and that's really a shame. They have now made the National Trust for Historic Preservation's "Most Endangered Historic Places" list with the 2001 addition of a barn built in Indiana in the 1850s.

The National Trust and *Successful Farming* magazine (with sponsoring support from Chevy Trucks) have joined forces to sponsor "Barn Again!" a national effort to help preserve our agricultural heritage by highlighting efforts and sharing how-to information on barn preservation. They provide information on everything from how to assess your structure's integrity to the tax incentives for undertaking such a preservation project. Some state preservation offices offer technical support, and some also offer grants or low-interest loans to help defray the cost of preserving barns and other agricultural structures. Check with your state preservation officer, and see resources, page 249, for more about "Barn Again!" and other preservation resources.

Alternative Structures

At one time, if you wanted a barn or chicken coop, you were limited in the kinds of materials you could use. Wood was the main material for years, but it has become prohibitively expensive as a construction material for many of us. Metal offers a reasonable alternative; it's cheaper and can contribute to an attractive, well-designed structure. Or you may want to consider some of the truly alternative approaches:

High-tech and low-cost plastics and fabrics are being used very successfully by many farmers to create hoop houses (a Quonset-shaped, greenhouse-like structure) and other unique solar structures. These structures are cost-effective (from initial cost to lower taxes compared to a traditional building) and work great for smaller producers. They go up quickly and easily, use natural ventilation, and contribute to fewer health problems in animals, but they tend to result in slightly slower animal growth during the coldest part of winter and the peak heat of summer, and they require a bit more labor for bedding and manure handling.

Straw was used as a building material in many Plains areas as early as the late 1800s, and today straw-bale construction is a rediscovered technique that is quickly evolving into a modern building system; it has been used successfully for barns and chicken coops around the country. Straw-bale construction can be economical if you have ready access to cheap straw, but if you have to purchase straw from far away, shipping costs can quickly eat up any cost advantage. Even without the advantage of cheap straw, however,

Hay Storage

If you will be storing hay in a mow, lean-to, or freestanding shed, allow for 250 cubic feet of space per ton of small, square bales or 310 cubic feet of space per ton of large round bales you expect to store.

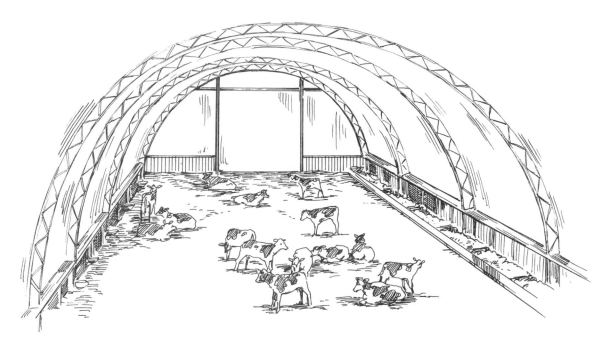

High-tech plastics are being used in "hoop houses." These structures have evolved out of commercial greenhouse technology and are gaining a strong following in the livestock industry; they provide excellent housing at reasonable cost.

Straw-bale construction techniques provide an economical alternative for animal housing projects.

the relative ease of construction and superior insulating capabilities (if properly ventilated, straw-bale buildings are warm in winter and cool in summer) make this alternative an attractive long-term approach.

Harvesting your own wood is another natural approach that can really reduce cost if you have access to forested land. Native lumber can be used successfully for most nonstructural wood (siding, floor, and roof sheeting), or you can dry it and then have it graded for structural use by a third-party certifier.

Using "small-diameter" logs is a new technique that is gaining attention and research. Historically, forests burned frequently, and small-diameter logs were cleared out in "cool" fires. Now we have millions of acres of forests that have been protected from fire for close to a hundred years; they are overgrown and therefore subject to large-scale, catastrophic fires. Most environmental groups agree that the forests need some thinning for their health but don't want to see the big trees removed. The answer is to find marketable uses for the small-diameter stuff, and small animal housing projects show promise of being a great use for these logs.

Building Kits and Packages

Kits and packages (some of which come as modular systems) have some advantages and some disadvantages: They are designed by architects and engineers to meet most building-code requirements, but some communities have stricter code requirements, so if you will be purchasing a kit or package, make sure the company will ensure compliance with codes where you live at the quoted price. All the supplies will be delivered to your site, cut to size, with the proper fasteners and directions for erecting the structure. Kits and packages may cost more than you would spend for a comparable structure built from scratch, but you are paying for convenience.

In spite of what a brochure may promise, opting for the kit or package route might not turn out all roses. We just built a kit storage shed; the instructions read like they were written by a lost tribe of Martians, and many of the predrilled pieces were drilled in the wrong place. Our shed was wood with an aluminum frame, and because we live in a particularly windy place, it was a serious challenge to get the frame up without the pieces bending out of shape. Next time we'll go back to building from scratch.

Although kits can be a do-it-yourself-type thing, a number of manufacturers of barn and stable kits work with specific contractors to put up their package for you, so all you do is write checks. Many of these packages are attractive and of high quality, but make sure you are dealing with a reputable firm. When considering one of these units, ask specifically for a detailed list of standard items and a list of optional items that cost extra. The brochure might show a building with beautiful big windows or nice wooden doors, but you find out after the fact that those are extras. Also, check with people in your area who have purchased a kit or package from a particular manufacturer, especially if you are buying something such as a complete barn.

When comparing packages, compare things such as warranties, delivery schedules and charges, and what recourse you have for solving problems.

You'll find contact information for some of the bigger kit/package manufacturers in the resources section at the back of the book.

Government Regulations

There are a few counties in the United States that have no building and zoning regulations, and a small percentage that have blanket exemptions for agricultural structures, but most likely you live in an area that has government regulations applying to all construction projects (with, at a minimum, electrical inspections required for any structure that has electricity run to it). Now, I realize almost everyone hates having to deal with these regulations, but remember, they were developed to protect health, safety, and property values for you, your neighbors, and future owners of your property.

Although some bureaucrats are challenging to work with, most are good people trying to do a tough job. Talk to them early in the process, when you are still making preliminary design decisions, and they will be supportive and helpful. Start building without having made contact, and they will be unhappy with you, which generally means increased cost and definitely means more time spent to finish a project. The primary concerns you'll need to discuss with your local building officials are zoning laws, building codes, and necessary building permits.

Zoning: Zoning laws are established by municipal or county governments to regulate land-use activities (such as designating zones for agricultural, residential, commercial, or industrial uses) and certain construction parameters (for example, minimum lot sizes, building heights, minimum or maximum building size, materials). Zoning laws can also regulate development in proximity to waterways and wetlands, and can define what kinds of animals you are allowed to keep on your land and how many. For example, a small-lot residential zone may preclude any livestock species or may allow a limited number of horses.

Building codes: Building codes are standardized design criteria that are developed by code-enforcement officers, engineers, and building-industry representatives. They are designed to ensure certain minimum standards for safety, but applied at their minimum, they don't necessarily result in efficiency or quality. Most communities adopt all or part of one of the "model codes" that are written by nationally recognized organizations, such as the Building Officials and Code Administrators International (BOCA), the International Conference of Building Officials (ICBO), the Southern Building Code Congress International, Inc. (SBCCI), and the Council of American Building Officials (CABO), though local governments are generally free to modify the model codes to meet their circumstances. The National Electrical Code (NEC) has been developed by the National Fire Protection Association, a nonprofit industry group representing the insurance community, fire departments, builders, and a host of other individuals who are involved in one aspect or another of building safety. Though it inherently has no weight of law, most communities or states have used the NEC as the basis of their electrical codes.

If you feel a code is going to prohibit you from doing something you really want to do, Paul R. Fisette, director of Building Materials and Wood Technology at the University of Massachusetts, Amherst, in his article "Decoding Building Codes," says,

Typically, inspectors will prohibit the use of materials and practices not prescribed by the code. However, inspectors are empowered to be flexible under the Alternate Materials and Systems section of the code. You may win an argument with an inspector involving a code interpretation if you can:

- demonstrate that your interpretation of the code follows the spirit and intent of the code;
- prove that your plan will deliver equally good results [you may need to employ a professional engineer or architect to verify that your plan will deliver good results];
- maintain a cooperative attitude during the negotiations. Solve the problem with the inspector, not in spite of the inspector. (http://www.umass.edu/ bmatwt/publications/articles/building_ codes.html [accessed Feb. 20, 2004].)

Building permits: A building permit is the tool municipal governments use to ensure your project conforms to zoning and building-code requirements. Most municipalities charge a fee for obtaining a permit. Many communities offer a waiver from building permits for structures that meet all of the following criteria: the structure is less than a maximum square footage (typically 120 square feet), it is not built on a permanent foundation, and it does not contain plumbing or electrical service.

Working with Contractors

If you opt to use contractors for all or part of your project, keep these thoughts in mind:

- It is best to prepare some "bid documents" that include plans and specifications and ask several contractors to bid on the job.
- Price is always important, but a cheap job done poorly is no bargain. If one bidder's price is much lower than those of other bidders, chances are he or she won't be able to do the job right for the price. Get clarification.
- Take your time and thoroughly evaluate prospective contractors. Check their refer-

ences. Questions to ask references:

— How was the contractor's overall workmanship?

— Did the job get done in a timely fashion?

— Did the contractor complete punch-list items? (A *punch list* is a list of small tasks to be done when the project is almost complete and may include items like touching up paint, repairing a faulty handle, and replacing a cracked window pane.) Payment schedules are generally written so that final payment is made when the punch-list walk-through is complete.

— Did the contractor willingly comply with the terms of the contract?

— Were warranties and guarantees honored?

• Inquire with the clerk of the local court to find out if the contractor has had significant lawsuits filed against the business, and with the county clerk to find out if the business has liens against it.

• Before signing a contract, confirm that the contractor is licensed (if required in your community), and that she or he has adequate insurance, including workman's compensation; public and property liability, which covers subcontractors as well as the contractor; and builders' general liability, which protects labor and materials on-site.

• Your own insurance agent should be able to provide guidance on what insurance and coverage limits the contractor should have in your state, and help you obtain a short-term rider for owner's protective liability, which will protect you in the event a liability claim arises out of the construction project.

• No work should be performed without a written contract between you and the contractor. Contracts should be absolutely clear on what the contractor does (supply all labor, materials, and equipment, or just some of the labor, materials, and equipment; a schedule with target dates) and on what you do (payment schedules, owner-supplied work or materials). Look for a "remedies" clause that

specifies what kind of damages can be sought if the project is not completed on time or if it fails to meet certain quality or contract conditions; but be aware that remedies clauses also entitle the contractor to seek damages against you for failure to meet your commitments.

• Most projects encounter some changes during the course of construction. Make sure the contract specifies how a change order (a written amendment to the contract that authorizes a change in work from the original contract documents) is initiated and how changes are priced. Change orders are generally required for price changes, material or equipment substitutions, changes of schedule, and for added work.

• You should specify that the contractor supply you with shop drawings, equipment and material warranties, instruction manuals, and any other items that come with things he is supplying you. Store these documents in a large, three-ring notebook. You can purchase plastic folders that go into three-ring notebooks that are excellent for storing booklets and plan sheets.

• Make sure your contract includes language that addresses the little things, like site cleanup and staging areas. To avoid the possibility of having a load of lumber dumped in the middle of your orchard, specify in writing where the crew can store equipment, park and drive heavy vehicles, or station a construction-site office trailer.

• Get everything in writing, even if you are working with a close relative: it will likely help you stay on good terms.

• Have an attorney review the contract for you. The bill at this phase is small compared to what you will pay an attorney down the road if things head for Kathmandu on an about-to-wreck train.

How Involved Will You Be?

There are three approaches to building projects, and which approach you opt to take will depend on the size of your project, the time you can commit, your level of experience with managing complex projects, and your finances.

1 *Do everything yourself.* This is the best approach for small projects, or if you have lots of time and are cash tight. Owner-builders can save anywhere from 20 to 40 percent, depending on how savvy they are, but if you have never before undertaken a construction project and you are considering a full-size barn, this is probably the wrong approach for you, regardless of potential savings. Start with something small — like a loafing shed or chicken coop — and if it goes well, venture for bigger things.

2 *Act as your own general contractor* but subcontract out some work. Contracting out pieces of work, like excavating, concrete work, and roofing, but doing other work, such as framing, electrical, and plumbing, yourself is the best approach. This gets a few critical pieces done quickly and by people who have the special tools and training to do the best possible job for those parts of the project. Savings range from 10 to 20 percent.

3 *Hire a general contractor* to take care of everything. For those with more money than time — or inclination — this is the best option.

My advice to do-it-yourselfers or those thinking of acting as their own general contractor: Know your limitations. I have seen many half-finished projects over the years, and in the long run they are more costly than just paying to get the job done quickly and well in the first place. Estimate time requirements and then multiply by four; estimate expenses and multiply by two. No matter which option you are considering, be a smart shopper: carefully compare materials and packages to ensure prices are based on apples-to-apples supplies; utilize competitive bidding techniques, including for the delivery of materials from lumberyards and national home stores (three bids is a good number); don't forget to factor in transportation costs and sales tax; and always check references.

What Constitutes an Independent Contractor?

The Internal Revenue Service has developed a comprehensive set of guidelines to clarify who is an independent contractor and who is an employee. As a general rule, if you have the right to control or direct only the result of the work done, then you are dealing with an independent contractor, but if you have the right to control the means and methods of accomplishing the result, then you have an employee, and you must withhold federal, state, and local taxes; you must pay unemployment insurance; and you must carry workman's compensation. The IRS takes a number of issues into account when determining independent contractor status, such as:

- What type of relationship do you have with the worker? Do you supply his or her tools and equipment? Do you specify tasks (for example, install the windows today), or do you specify a finished product (a building framed with windows, doors, and roof installed)? Do you set the workers schedule of days and/or hours?
- Who hires additional help? Who pays the additional help?
- How did the worker obtain the job (through an application, a bid, an employment agency)?
- Is there a contract in writing between you and the worker? Does the worker have contracts with others? Does the worker advertise his or her services to others?
- Do you supply the worker with benefits, like vacation or holiday pay and health insurance?
- What types of expenses does the worker incur? Does he or she have to be licensed and insured? Does he or she have to purchase tools and equipment? Does he or she have to supply materials to complete the project?
- What type of payment does the worker receive — hourly wage, salary, lump sum, piecework?

Ernie LaBombard

Dreaming of a truly unique barn with lots of character? Ernie LaBombard may be able to help. Ernie started working in post-and-beam construction in 1973. At first he was building new houses using this age-old technique, but, as he says, "Somewhere along the line things got kind of slow workwise and we stopped building new houses. Then somebody asked me to put up an antique barn that was in pieces, and I said, 'Yeah, I can try that.' "

At that point Ernie's brother, Jessie, started working with him, and they reconstructed that old barn in Hanover Center, New Hampshire. That started them on the path of "doing barns," and the rest is, as the saying goes, history. Today the brothers run a thriving business called Great Northern Barns that buys antique barns about to be torn down and resells them, handling everything from the removal to the refurbishment and installation on a new site.

The process is fairly expensive, so most of these grand old dames are adapted for other uses, such as houses or commercial structures, but Ernie says at least a few each year are rebuilt to play their traditional role as a barn.

Ernie explains that there are a lot of nice old barns out there that need saving. "Some of these barns are 200 years old, or older, and if somebody doesn't take them and use them as a new structure, then they just get lost; the roof goes and they fall down, or they get torn down to make way for a road expansion, or for construction of a shopping mall. Every year we lose hundreds of them around the country, so if somebody can take one and save it . . . well, it's like saving an orphan. It's a neat thing to do."

The best place to find an antique barn for restoration is in an area that has really good farmland and was settled early. Although the Connecticut River Valley, where Ernie and Jessie work, is just such a place (it has "a treasure trove of antique barns"), other areas of the country also offer opportunity for those seeking to save an orphan.

According to Ernie, if you are shopping for an antique barn, look

> " . . . it's like saving an orphan."

for older ones that were hand hewn and haven't been changed much over the years. If the owner has kept a good roof over it and it's still standing, then it's probably suitable. But he also recommends that you "get somebody who knows timber-frame construction or a 'structural' person to come and look at things, because the small stuff — the trim, the roofing, the siding, the windows — isn't really very important compared to the structure. Even buildings that are still fairly sound almost always need something fixed on them. The average 'frame' probably needs a dozen serious fixes, like the repair of a major joint, or a nailer that is just totally gone and needs to be replaced."

A crew like Ernie's can take down a barn in a week. They'll spend about a month on cleaning the structure and doing repairs in the shop, and then they can get it back up in about a week.

3. STRUCTURE, DESIGN & FUNCTION

Several structural, design, and functional issues must be considered carefully during the planning phase before you select a design and begin to finalize your plans. Some of the considerations are aesthetic, like the kind of roofline you want and the styles of doors and windows you'll use, but others are important to a building's performance, such as the type of foundation and the best type of ventilation for your particular needs. The brief overviews in this chapter are intended to help you with planning and decision making; for a more in-depth discussion of these topics, see part III, Construction.

FOUNDATIONS

Your building will only be as good as its foundation. The foundation is designed to disperse the building's load over a large area of subsoil and to secure the building to the ground.

Small structures such as a chicken coop or rabbit shed can be built on skids so they can be moved, but larger, permanent structures need to be connected firmly to the ground. There are several types of foundations generally used in animal housing — each with advantages and disadvantages. They are:

Skids. Created from 6x6 or 8x8 timbers, skids can be semipermanent when placed on gravel-filled trenches. For portable structures, skids are made from 4x4 timbers, with the edges trimmed for ease of movement. Semipermanent skid structures can be one story tall and occupy an area of up to about 144 square feet. Although a semipermanent structure can be moved occasionally, it is not designed for routine moving and is best set on a rock or gravel area that provides drainage. The portable structures are designed for routine movement in order to take advantage of pastures or other homegrown feed sources. They are short structures, typically less than a full story tall, and usually smaller than 60 square feet.

Slabs. Poured concrete slabs are created with reinforced wire and are 4 to 8 inches

TYPES OF FOUNDATIONS

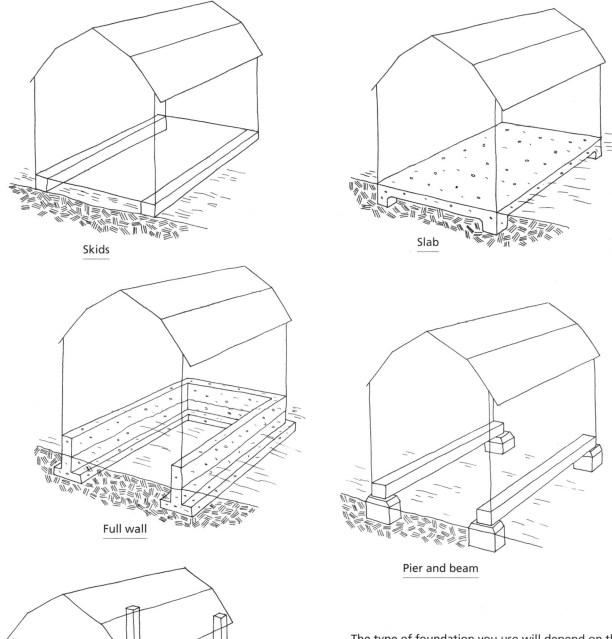

Skids

Slab

Full wall

Pier and beam

Pole

The type of foundation you use will depend on the size and use of the building. Skids work well for small buildings, particularly semipermanent or portable structures. Slabs are often used in areas of the barn that will house offices, storage, or other special-use areas, but are hard on animal's legs. Full-wall foundations are rare in most modern animal housing projects, but they may be necessary for large, two-story barns. Pier and beam, and poles on footers are seen more commonly thanks to ease of construction and price.

thick. Typically the slab and foundation are cast as one piece that is laid on top of concrete footings. Slabs are economical and good for many animal housing projects where you want a solid floor. They are required in milking rooms and dairy parlors. If animals will be walking directly on concrete slabs, the surface needs to be roughened with grooves to provide traction. If not well supported on good footers (that reach below the frost line or are built using frost-protected shallow foundation [FPSF] techniques that incorporate insulation to reduce the depth of the footer), slabs are susceptible to cracking and chipping.

Walls. Wall foundations are generally either poured concrete or laid-up block walls that are built on footers (though in areas of Alaska and Canada, they may be built with specially prepared wood, because using concrete is so challenging). Wall foundations yield a crawl space or basement, which makes them unusual in animal housing projects; the benefit of a crawl space or basement generally doesn't justify the additional cost.

Piers or pilings. Piers and pilings are constructed of poured concrete (or pressure-treated wood, wood encased in concrete, or steel) and are bridged by strong beams. They are economical and common in midsize projects.

Post. Post foundations are probably the most common approach for larger animal housing projects because of their economy; they provide a foundation and a wall system at the same time. Large wooden posts or steel beams are set below the frost line on a poured footer, and the building is constructed directly on the posts. Typically these structures have gravel floors, sometimes in combination with slabs, wooden floors, or paving stones, particularly in offices and tack or feed rooms that are included in the structure. In the past, these structures were referred to as *pole barns,* and they were constructed with poles similar to telephone poles, but now the industry prima-

rily uses square posts, which make construction quicker and easier.

FLOORS

Floors don't have to be elaborate when you are building animal housing. Often, bare-earth floors are all that is needed, with a thin layer of gravel added to improve drainage. This is an economical option that provides secure footing for animals and reduces stress on their legs.

In the damp environment of a barn or stable, wooden floors are not the best choice; they rot out fairly quickly from urine and manure. Wooden floors do work well in storage areas, offices, and other nonanimal areas, however. Small sheds on skids often incorporate plywood floors. You can extend the life expectancy of such floors by using marine-grade plywood.

Durability isn't a concern with concrete or paving-stone floors, but they are hard on animals' legs. If animals will spend a great deal of time in stalls on these hard surfaces, consider using bedding or rubber mats designed for barn use to reduce stress on their legs.

Bedding

Bedding, or litter, is used to absorb moisture from urine and manure and to provide comfortable areas for animals to rest. Bedding needs to be cleaned thoroughly on a daily basis or, alternatively, a technique known as deep bedding can be used.

In the deep-bedding system, a 6-inch layer of bedding is placed on the ground. For small animals such as birds and rabbits, manure and wet spots are cleaned off the surface of the bedding once a week; for large animals, they are cleaned off every day or two. After removing the soiled areas, new bedding is added — an inch or two is generally plenty — so the area is again dry and clean.

The bedding pack builds up over time, so once every year or two, it must be cleaned out

Deep bedding provides a clean and comfortable environment. Build up an initial 6-inch layer of bedding (straw, shredded newspaper, wood shavings, dried leaves). Depending on animal size, you'll clear the surface of manure and wet spots every few days or weekly, then add a little dry bedding.

completely. When you clean out the pack, you will discover partially composted material. Pile it into a compost pile, and a year later you will have the finest soil amendment in the universe. Cleaning bedding is easiest if you design your building to accommodate a tractor equipped with a loader bucket, but it can also be done gradually by hand.

Deep bedding not only yields great compost for the garden, but it also provides animals with a clean, dry area that is warm and well cushioned. And there are fewer ammonia fumes with deep bedding because the pack absorbs the urine.

WALL SYSTEMS

Decisions about wall systems affect the cost and complexity of a building as well as its appearance. Because post-frame construction is the most economical approach to barn building, it is the most common wall system used for larger barns today; timber-frame, stud-wall, and poured concrete are other options.

Post-frame is one of the most common types of animal-housing construction systems, because it offers a structurally sound design that's cost-effective, relatively simple to construct, and quick for framing up a structure. In post-frame construction, posts provide the foundation and support bearing walls, partition walls, the roof, and the floor. The technique provides great flexibility in size; it can be used for a 10-foot by 10-foot shed and for buildings the size of a football field. However, post-frame construction is not usually suitable for two-story structures. (See illustration on page 116.)

Timber-frame construction is a traditional style of construction also known as post-and-beam construction. It typically has a concrete foundation, and its structure is held together not by nails, screws, or bolts but by wooden dovetails and mortise/tenon joints. Timber-frame buildings are beautiful but expensive.

Amish barn builders still use the technique, and it is used in specialty applications like heritage farms, but for the average person wanting to build a barn, timber-frame construction is probably out of the realm of possibility.

Stud-wall construction, or "stick built," is fairly common for small farm buildings such as sheds and chicken coops (it is also the most common construction technique for single-family housing), but because of cost, it is unusual for larger barns. It uses 2x4 or 2x6 studs to frame up walls and is typically supported on a concrete foundation.

Poured-concrete and *concrete-block walls* are not too common in most of the country, but they are often used in desert areas because they help keep the interior of the building cool. They are also popular for small dairy facilities because they can be kept clean with regular power washing. They are not expensive and can be formed, or "laid up," quickly by an experienced mason.

THE ROOFLINE

The roof ties your structure together, and it also plays an important role in establishing the look of your building. A variety of roof styles can be used for animal buildings. A roof style depends somewhat on your tastes and may reflect regional preferences and trends, but it also depends on the type of structure it will cover.

The easiest roof to build and cover — with roofing tin or shingles — is the shed roof. It slopes in only one direction, and it is well suited to three-sided sheds and any other kind of small structure. The gable style (which slopes in two directions) and hip style (which has slopes in three or four directions) are also easy to build and are common throughout the United States. A number of roof designs, like the saltbox (different slopes in two directions), are adaptations of one or more of these styles.

Roof types commonly seen on barns include the gambrel (the typical dairy-barn

style), the gothic arch (seen primarily in the East), the mansard (more common in the Northeast, particularly areas that had an early French influence), and the monitor (often seen on horse barns and fairly common in the West); these types are more complicated and expensive to build, but they have some advantages, too. For example, the gambrel and gothic roof styles provide large areas on the second floor that can be used for hay storage, which was their traditional purpose. But loading hay into a second-story mow is hard work, believe me! We know of people who have installed offices, studios, and guest quarters in these second-story spaces, and they're great for that purpose.

LOADS

Our friends Theresa and Dennis built a beautiful little barn. They live in a county that exempts agricultural buildings from the building code, so they designed and built it themselves without having the plans reviewed by anyone. They didn't know about structural loads, and shortly after they completed their barn, it collapsed when they put too much feed in it. They worked hard to get it back up, but a strong wind came along and, due to its already compromised state, lifted it and splintered the building. The barn is now unsalvageable. Theirs was a cruel lesson.

Engineers, architects, and building department officers all learn about structural analysis for loads. They design the building, or review designs, according to the type of loads the building will be required to sustain. (To avoid the type of trouble that Theresa and Dennis suffered, it is critical to involve these folks, at a minimum, in a review of the plans for structural loading.)

Loads are often divided into two classes: dead loads, which are constant in magnitude and location throughout the building's usable life, and live loads, which vary over time and place. Dead loads are generally associated

TYPES OF ROOFS

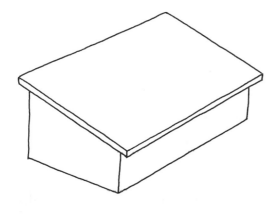

Shed

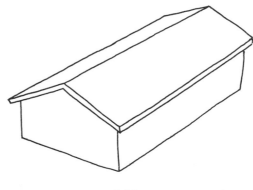

Gable

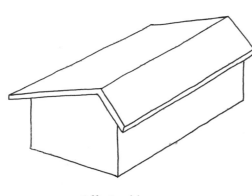

Offset gable

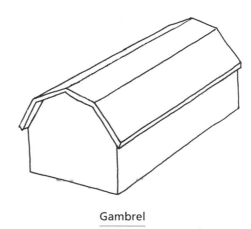

Gambrel

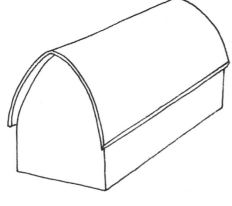

Gothic

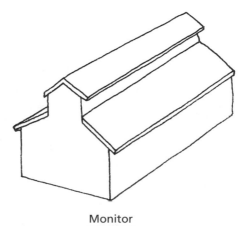

Monitor

Roof designs often reflect aesthetic considerations and regional preferences. Shed roofs are easiest to construct, followed by gable roofs. Other rooflines are more complicated and more expensive to build, but they can provide extra space in two-story structures.

STANDARD LOADS	
Item	Loads (lbs./sq. ft.)
Asphalt shingles	2.00
Copper, tin	1.00
Metal, 20 gauge	2.50
Metal, 18 gauge	3.00
Wood shingles	3.00
Corrugated cement roofing	4.00
Wood sheathing (per inch of thickness)	3.00
Skylight (metal frame, ⅜-inch glass)	8.00
Plywood (per ⅛ inch of thickness)	0.40
Gypsum board (per ⅛ inch of thickness)	0.55
Fiberboard (½ inch)	0.75
Lath and plaster	8.00
Wood furring suspended ceilings	2.50
Fiberglass insulation (per inch of thickness)	0.70
Cattle or horses*	100.00
Sheep or goats*	40.00
Swine (to 400 pounds)*	65.00
Poultry*	20.00
Shops, maintenance areas, milk rooms	70.00
Hay or grain storage	300.00
Farm tractors	200.00
Light trucks	100.00
Snow loads (depending on area of country)	0.00–150.00
Wind loads (depending on area of country)	10.00–85.00

*Increase loads by 25 percent in areas where animals will be crowded.

with gravity working on the building's materials; in other words, it is determined by the weight of the roofing, flooring, walls, and permanently installed equipment. Live loads change over time and are associated with the weight of the people and animals, equipment, feed, tack, and other materials inside the building, and with the pressures applied by wind, snow, or earthquakes. Wind, snow, and earthquake loads can be the most variable, depending on where you live, but your local building official can provide design criteria that address these concerns, even if your building will be exempt from code enforcement.

Load-bearing components of the building, such as the roof, exterior walls, columns, and floors, must be able to sustain the weight of all the loads that might be applied to them. When they are unable to support the loads, the building system fails, resulting in problems ranging from cracking and shifting foundations to complete collapse.

DOORS AND WINDOWS

During the planning period, give thought to where doors and windows will be placed and to their size and style. Doors must be wide enough to accommodate whatever will be moved in and out of a building. Careful thought about the placement of doors and windows can make your structure user-friendly and enhance natural ventilation. A Dutch door, for example, can double as a source of light and air, while placing windows high on a wall takes advantage of winter sun and helps reduce summer heat. (See pages 218 and 221 for illustrations of door and window types.) Doors that tractors and other farm equipment will pass through should be at least 10 feet wide, though 12-foot doors are even better; obviously, the doors must be tall enough to accommodate the tractor's cab or rollover protection bars. Doors for large animals, like horses or cattle, should be at least 4 feet wide, and small animals such as sheep and goats require at least 3 feet, which is the width of a standard-sized exterior house door. Doors should be easy for you to open with one gloved hand but should not be easy for animals to open.

Windows need to be protected from animals by heavy screens or some type of grill-work, although the screening or grill can make opening and closing the window a challenge. Another option is to place windows above an animal's reach, hinged so they open in from the top, which improves ventilation without creating a draft. Place most of your windows on the south side of a building to take advantage of winter solar energy and summer breezes.

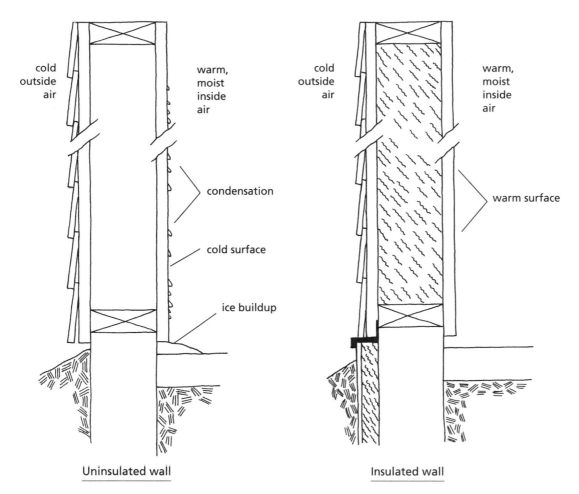

cold outside air

warm, moist inside air

condensation

cold surface

ice buildup

Uninsulated wall

cold outside air

warm, moist inside air

warm surface

Insulated wall

Insulation can reduce heat exchange between the outside and the inside environment, keeping a building warmer in winter and cooler in summer. Uninsulated structures will have problems with condensation and will have ice buildup on the floor near the base of outside walls. Insulated walls reduce these problems, keeping the structure sound and your animals happy.

INSULATION

Insulation works by reducing heat transfer by conductance and convection, thereby keeping buildings warmer in winter and cooler in summer. Even open buildings like a three-sided loafing shed can benefit from having a layer of insulation added to the ceiling: it will reduce summer heat and condensation in winter.

Use a vapor barrier on the interior side of insulated walls. Six-mil polyethylene film is usually the best choice. Taping joints, nail holes, gaps around electrical fixtures, and other holes in the barrier improves performance.

BALANCING TEMPERATURE AND HUMIDITY

As we learned in chapter 1, air quality is critical in animal health. In the course of normal breathing, animals give off moisture as they exhale. When temperatures rise, they begin to pant, thereby increasing the moisture they give off. A mature dairy cow, for example, gives off about 0.85 pound of water per hour at 30°F, but when the temperature is 80°F, she more than doubles that, at 1.98 pounds of water per hour. In an enclosed building, this moisture is captured in the air, creating high humidity. As the temperature

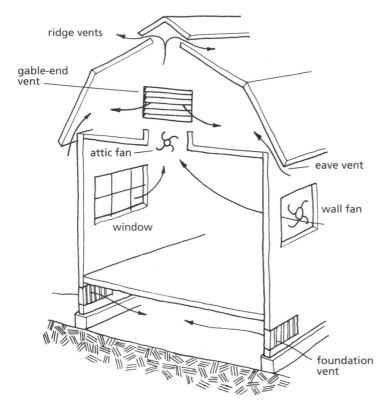

Ventilation is critical to animal health. Provide healthy air for animals with mechanical ventilation (fans) or natural ventilation (windows; ridge, eave, or gable-end vents).

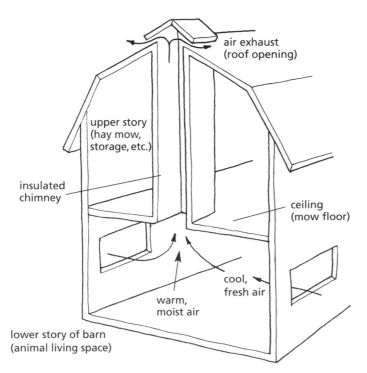

Chimneys aid in natural ventilation in two-story barns by providing a passageway for hot air to move up from the first floor and out roof vents or cupolas.

of the air increases, it is able to hold more moisture. When both temperature and humidity are high, an animal will suffer from heat stress, which at a minimum reduces production (meat, milk, fiber, eggs) but can also result in death.

The ideal humidity level in a barn is in the range of 40 to 60 percent. This range helps keep down dust, and it also minimizes airborne pathogens and condensation. Condensation forms where warm, moist air comes in contact with a cool surface, such as a wall or ceiling, and it can cause buildings to rot.

Air entering a building is generally cooler than inside air (even on a hot day), so it removes moisture and cools the inside air. During cold weather, airflow needs to be sufficient to remove excess humidity, and in hot weather it must also reduce temperature. Consequently, ventilation systems need to be designed to achieve a balance between temperature and humidity throughout all seasonal changes — hot, mild, and cold weather. Two-story barns require extra attention to ensure adequate ventilation, and they should include flues or chimneys to move stale air from the lower level up to a ridge vent or cupola.

HEATING

Three things affect indoor temperature: the ambient air temperature outdoors, body heat given off by animals, and heat generated by lights or equipment such as compressors. Although mature animals don't generally require a heated barn, baby animals may if they're born in winter. Supplemental heat may also be called for in milk rooms and parlors, holding areas for sick animals, or in areas that you will be spending time in, like an office, tack room, or shop.

In a small or multipurpose barn, radiant heaters are often used to supply supplemental heat for baby animals. These heaters can be infrared heat lamps or gas units placed over a pen or enclosure. They must be hung high enough above the pen not to be a fire hazard or threat to the animals. No more than six 250-watt heat lamps should be used on a 15- or 20-amp breaker. Although piglets and chicks (poults, ducklings) require supplemental heat for an extended time, lambs, kids, calves, and foals need it only immediately after birth if at all; once they are dry, supplemental heat can be turned off.

For specialty barns that will be regularly used for babies — like a pig-farrowing barn and nursery — heat is often supplied through radiant floor systems. These can be con-structed in the floor using hot-water piping, hot-air piping, or electrical resistance cables. Another approach is to use fiberglass mats with coils in them placed on the floor.

Area heating for a milk room, for example, is usually supplied by space heaters. Vented gas heaters are the best choice for routine use, because they are economical and safe. Unvented gas heaters should not be used, because they exacerbate any problems with gases like carbon monoxide, and electrical space heaters are expensive if they are run with any regularity.

ACCEPTABLE TEMPERATURE RANGES FOR LIVESTOCK

Animal	Minimum temperature in °F (°C)	Maximum temperature in °F (°C)
Cattle	40 (4)	60 (16)
Calves	50 (10)	70 (21)
Horses	40 (4)	60 (16)
Foals	50 (10)	70 (21)
Sheep, shorn	40 (4)	60 (16)
Sheep, unshorn, or goats	30 (–1)	50 (10)
Lambs or kids	50 (10)	75 (24)
Pigs	50 (10)	70 (21)
Pigs, 50–75 lbs.	60 (16)	70 (21)
Pigs, 3 weeks to 30 lbs.	75 (24)	85 (29)
Pigs, newborn to 3 weeks	85 (29)	95 (35)
Hens, layers	55 (13)	70 (21)
Broilers, turkeys, other growing birds	65 (18)	70 (21)
Chicks, poults, other birds in brooders	70 (21)	95 (35)
Rabbits	40 (4)	60 (16)

Note: With all baby animals, the ideal is to start them at the high end of the range and move them over the course of weeks to the low end of their range.

Dr. David Kammel

"How are the animals going to breathe?"

Since 1985, Dr. David Kammel has been the state Extension and research specialist for livestock housing at the University of Wisconsin. As well as teaching some short courses on livestock housing, his job requires him to regularly field questions about livestock housing issues from county Extension agents, equipment suppliers, and farmers and animal owners. Although occasionally the questions come early in the planning process, more often than not they come after construction plans have been selected and paid for and even after a building has been built. Problems associated with livestock housing often result in increased veterinary bills.

Dr. Kammel says that when people call early in the process, he encourages them to "write stuff down on paper! It might seem like a silly, simple thing to do, but invariably nobody does it. People are trying to work on a decision, but they don't have any basis for that decision because they haven't written anything down.

"The first questions I ask them to answer on paper are about scale: 'How many and what kind of animals do you have now and what's the future size of the farm, or how many and what kind of animals might you have in the future?' "

He also encourages people to list, in two separate columns, things they *want* and things they *need*. That is, "You need these things for keeping these animals healthy — to feed them, to handle manure from them [etc.], and then there are things that you want but that aren't necessarily what the animals need, like an office or a laundry room. When you begin budgeting, the list allows you to control costs by cutting from the *wants*, not the *needs*. The bells and whistles might be really cool, but you have to ask, 'Are they really going to serve the purpose?' "

As Dr. Kammel points out: "You don't build a house for your family size of two now, but for how many children you might have in the future. If you can't afford to build it all right now, by taking time to do good planning and design, you can build a structure that will be relatively easy to add on to, and you are not going to waste money to build an infrastructure that you can't use down the road. I don't think it matters what size you are starting at — it could be just a few cows or horses now, but someday it might be four or forty animals."

The biggest fault that Dr. Kammel sees in plans and buildings is insufficient consideration of ventilation. Even with the package barns that are commercially available, Dr. Kammel says, "more often than not they are *not* properly ventilated. Ask the question when you're planning your animal housing, 'How are the animals going to breathe?'

"To put it simply, you have to ensure that you can control how much fresh air is going in and how to get the stale air and dust out; then figure out what it's going to cost and decide how you're going to pay for it. Ventilation can be natural or mechanical, but it has to be designed to match the structure."

According to Dr. Kammel, horse barns are notorious for being poorly ventilated; the reason is "what we think a horse needs isn't what a horse needs; it's what a person needs." He adds, "When it is all said and done, if you are looking at trying to keep a horse as healthy as it can be, then a run-in shed on a piece of pasture is the best place for that horse!"

II. PLANS

Years ago, Ken and I both ran municipal wastewater treatment plants, and each of us spent time as the lead employee for major expansions at these types of facilities. The plans and specification packages (*specs* for short) for those projects were huge, with hundreds of pages of plans (also called *drawings*) and as many as five 3-inch-thick books of specs. Luckily, an animal housing project won't require such extensive plans and specs, but it will nonetheless require some.

The combination of plans and specs become part of any contract documents you draw up with contractors. Plans are prepared to some consistent scale and typically include:

• A *perspective drawing* that shows the building from an angle; this is meant to make the plans easier to understand for the homeowner. As such, perspective drawings are not required in formal plans that are presented to contractors, building officials, and other professionals.

• A *floor plan*, which shows the general layout of the building, including interior walls, doors, windows, and special features like stalls, bathrooms, and storage areas.

• *Elevations*, which show each side of the building on a single plane. Elevations may be difficult to understand at first because everything is shown in two dimensions.

• *Sections and detail plans*, which show the elements of a structure in great detail. For example, these drawings might show how a ventilation system or insulation is installed, or the foundation in detail. (Sections are called out on floor plans and elevations with bracketed arrows labeled with letters.)

• A *site, or plot, plan*, which shows where the structure will be situated and its relation to other features, such as lot lines, buildings, roads, easements, waterways, and so on.

The specs provide additional written guidance for the construction project. They define in writing what work is to be done, products to be used, standards to be met, and guarantees to be included.

Plans and specs can be developed by an architect or engineer, purchased from plan-service companies or government organizations, or you may be able to develop them yourself based on the plans that follow. The plans in this chapter are intended to provide inspiration and ideas, but they'll probably require modification for your location or particular application. It's wise to have plans reviewed by a professional to assess structural loads *before* construction begins. For sources of plans and for suppliers of commercial products, see the resources on page 249.

Keep Organized

Before you start a project, get a three-ring binder and store documentation, warranties, receipts, brochures, and all other pertinent information in it as the project goes forward. Keep a set of "final" drawings that show precisely what has been done, such as the exact locations of buried lines, electrical junctions, and so on. The final drawings can be folded and inserted in the binder in plastic envelopes designed for that purpose. When the project is finished, all the information about it will be collected in the binder. This is handy if you need a warranty or to calculate the value of the improvements, which can offset capital gains if you sell your property.

4. SMALL & PORTABLE HOUSING PROJECTS

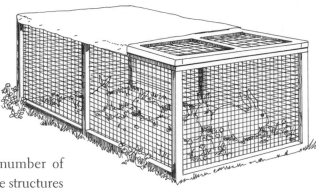

This chapter covers plans for a number of small and portable structures. These structures are great for learning building techniques, but they are also very practical for keeping animals. For most small projects, it is easiest to cut all the pieces first and then begin assembly. Small nails or screws work well, and using wood glue on all seams and connection points helps to reduce moisture seepage and prolong the life of the structure. Those that are left natural or painted a light color on the outside will be cooler inside (desirable in most circumstances) than those painted a dark color.

RABBITS

One of our first animal adventures, back in the early 1980s, was raising rabbits for meat and to sell as pets. We started with a couple of bunnies in a dog kennel and added some ducks to it. They shared a small structure built of stacked concrete blocks, with a sheet of farmer's tin (galvanized metal siding) for a roof. Later, we remodeled an old coal shed into our "barn," in which we included stacked rabbit hutches, ducks and geese, and one wether goat that had been abandoned in town.

Although our first attempts at raising rabbits in cages were successful, we learned that rabbits benefit from the chance to graze on fresh green grass; an ark, like the one shown above, provides ample opportunity for this. Arks can serve as principal rabbit housing, but they may need to be protected by a portable electric fence, especially if the rabbits will be in them at night when predators, including the neighbors' dogs, are most active. Alternatively, arks can be used in combination with permanent hutches, or cages that are installed in secured buildings. The bottom of a rabbit ark should have chicken wire stapled to it so the rabbits can't burrow out. By stapling a piece of canvas (or plastic tarp) over half the top, you provide protective shade from the heat of the day.

Permanent rabbit hutches are heavy-duty cages; they work well for a small rabbit endeavor (up to four adults), but for larger operations, try a rabbit shed with wire cages. A small shed can accommodate six to twelve cages, which can be hung on wires so manure drops to the ground. This is a clean and easy approach for rabbit raising. The wire cages can be constructed by hand with 2x2 lumber and hardware cloth or purchased as a wire-only cage (premade or as a kit).

Author's note: In this chapter, I refer to *arks,* which in this context are portable structures that provide shelter and a small pen for grazing.

Rabbit Hutch

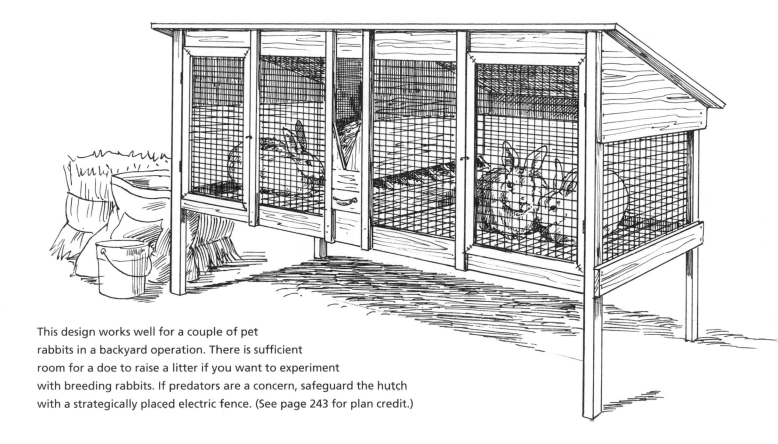

This design works well for a couple of pet
rabbits in a backyard operation. There is sufficient
room for a doe to raise a litter if you want to experiment
with breeding rabbits. If predators are a concern, safeguard the hutch
with a strategically placed electric fence. (See page 243 for plan credit.)

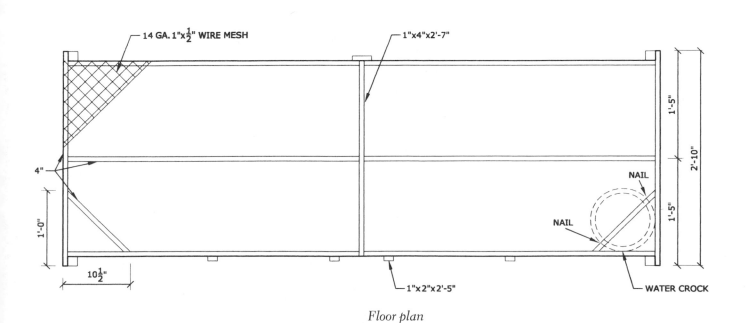

Floor plan

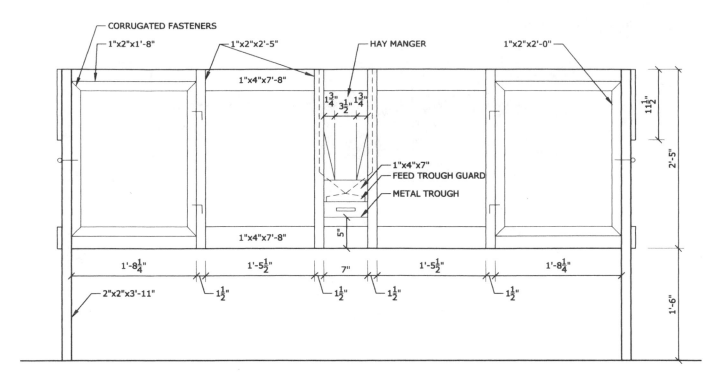

Front elevation

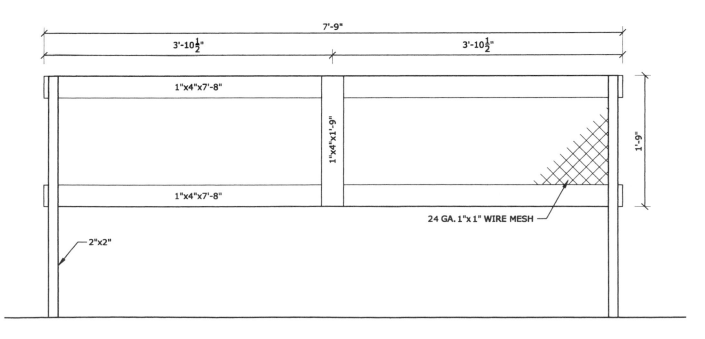

Rear elevation

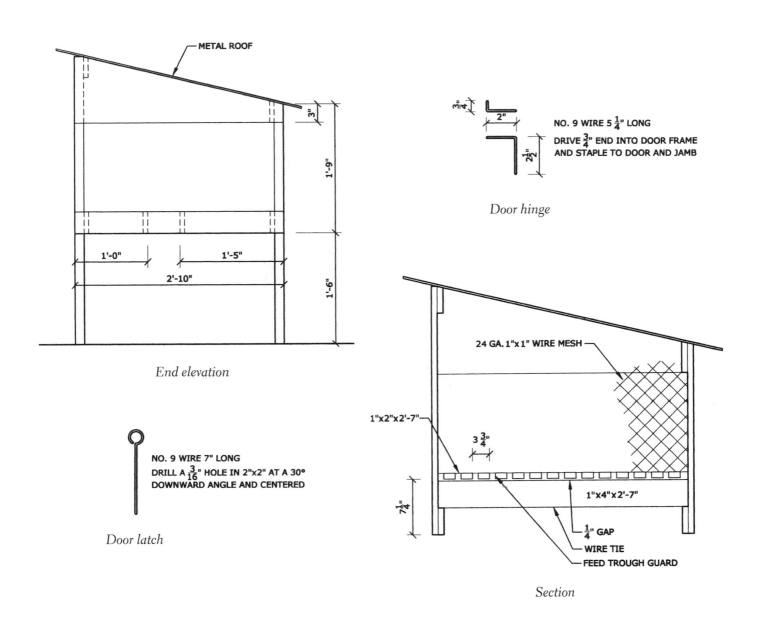

MssETAL ROOF

3"

1'-9"

1'-0"   1'-5"

2'-10"

1'-6"

End elevation

NO. 9 WIRE 7" LONG
DRILL A $\frac{3}{16}$" HOLE IN 2"x2" AT A 30°
DOWNWARD ANGLE AND CENTERED

Door latch

$\frac{3}{4}$"

2"

NO. 9 WIRE 5$\frac{1}{4}$" LONG

$2\frac{1}{2}$"

DRIVE $\frac{3}{4}$" END INTO DOOR FRAME
AND STAPLE TO DOOR AND JAMB

Door hinge

24 GA. 1"x1" WIRE MESH

1"x2"x2'-7"

$3\frac{3}{4}$"

$7\frac{1}{4}$"

1"x4"x2'-7"

$\frac{1}{4}$" GAP

WIRE TIE

FEED TROUGH GUARD

Section

GA. = gauge

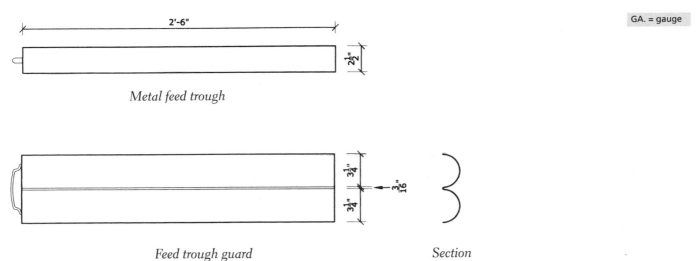

2'-6"

$2\frac{1}{2}$"

Metal feed trough

$3\frac{1}{4}$"

$\frac{3}{16}$"

$3\frac{1}{4}$"

Feed trough guard   *Section*

Two-Compartment Rounded-Corner Hutch

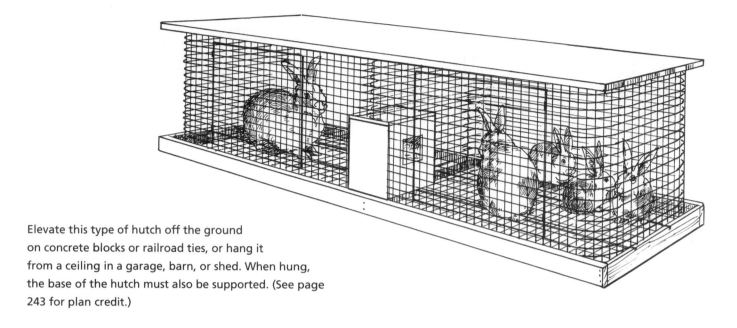

Elevate this type of hutch off the ground
on concrete blocks or railroad ties, or hang it
from a ceiling in a garage, barn, or shed. When hung,
the base of the hutch must also be supported. (See page
243 for plan credit.)

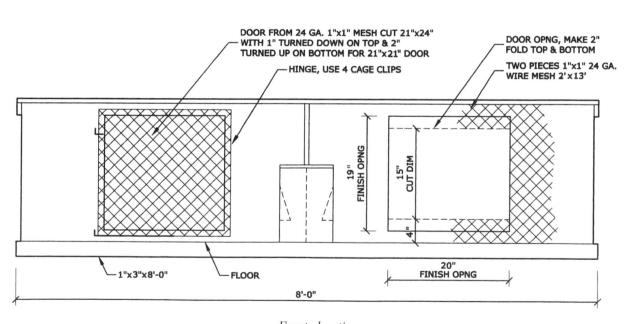

DOOR FROM 24 GA. 1"x1" MESH CUT 21"x24"
WITH 1" TURNED DOWN ON TOP & 2"
TURNED UP ON BOTTOM FOR 21"x21" DOOR

HINGE, USE 4 CAGE CLIPS

DOOR OPNG, MAKE 2"
FOLD TOP & BOTTOM

TWO PIECES 1"x1" 24 GA.
WIRE MESH 2'x13'

19" FINISH OPNG

15" CUT DIM

4"

1"x3"x8'-0"

FLOOR

20" FINISH OPNG

8'-0"

Front elevation

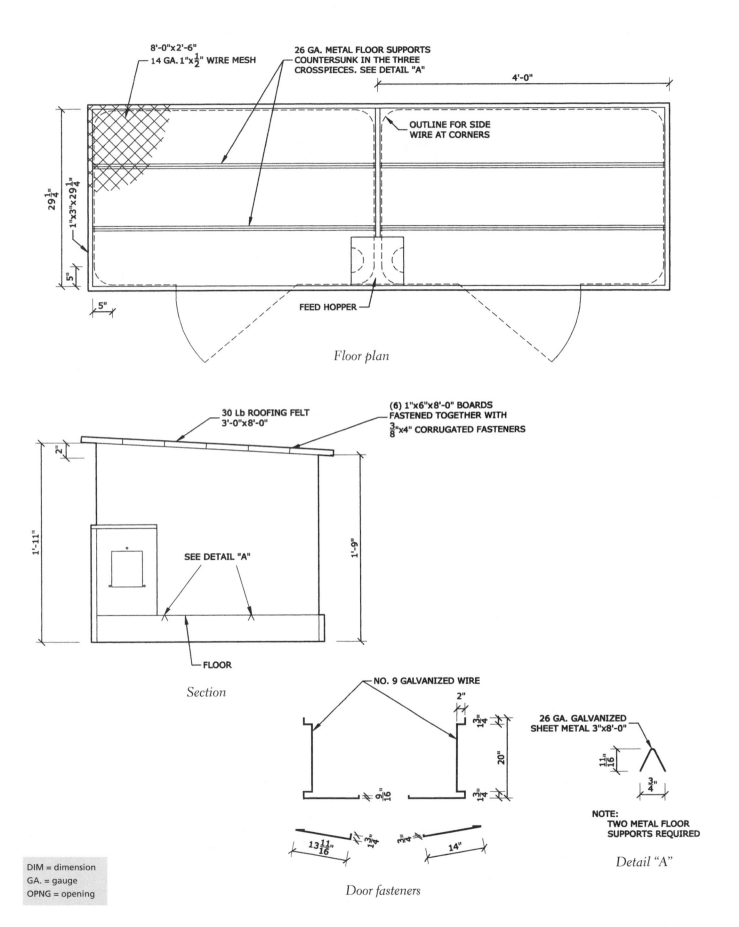

8'-0"x2'-6"
14 GA. 1"x½" WIRE MESH

26 GA. METAL FLOOR SUPPORTS
COUNTERSUNK IN THE THREE
CROSSPIECES. SEE DETAIL "A"

4'-0"

OUTLINE FOR SIDE
WIRE AT CORNERS

29¼"

1"x3"x29¼"

5"

5"

FEED HOPPER

Floor plan

30 Lb ROOFING FELT
3'-0"x8'-0"

(6) 1"x6"x8'-0" BOARDS
FASTENED TOGETHER WITH
⅜"x4" CORRUGATED FASTENERS

2"

1'-11"

1'-9"

SEE DETAIL "A"

FLOOR

Section

NO. 9 GALVANIZED WIRE

2"

1¾"

26 GA. GALVANIZED
SHEET METAL 3"x8'-0"

20"

11"/16

1¾"

9"/16

¾"

NOTE:
TWO METAL FLOOR
SUPPORTS REQUIRED

Detail "A"

13 11/16"

1¾"

¾"

14"

Door fasteners

DIM = dimension
GA. = gauge
OPNG = opening

Rabbit House

A small rabbit house is ideal for anyone who wants to breed rabbits for meat or to sell as pets or for 4-H projects. This house can accommodate as many twelve cages, or it can do double-duty, housing rabbits and chickens or rabbits and tools or garden supplies. (See page 243 for plan credit.)

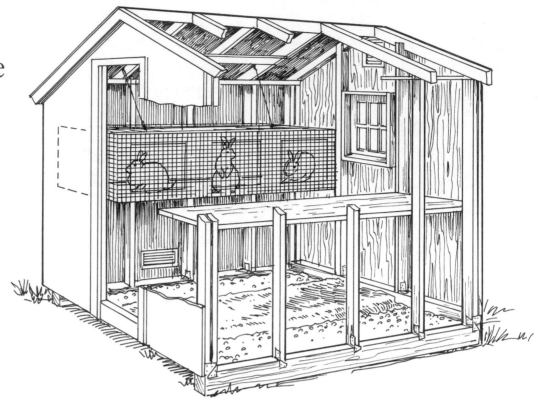

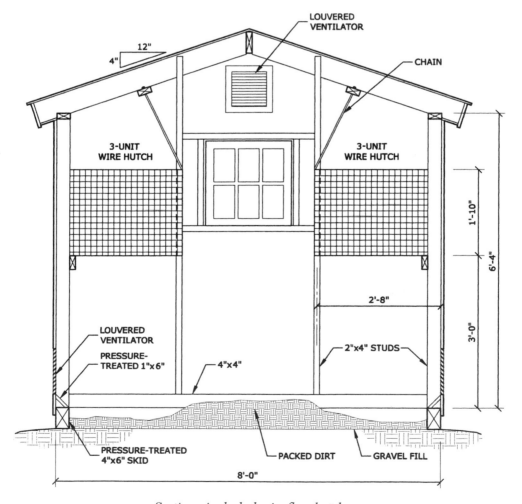

Section: single-deck wire-floor hutch

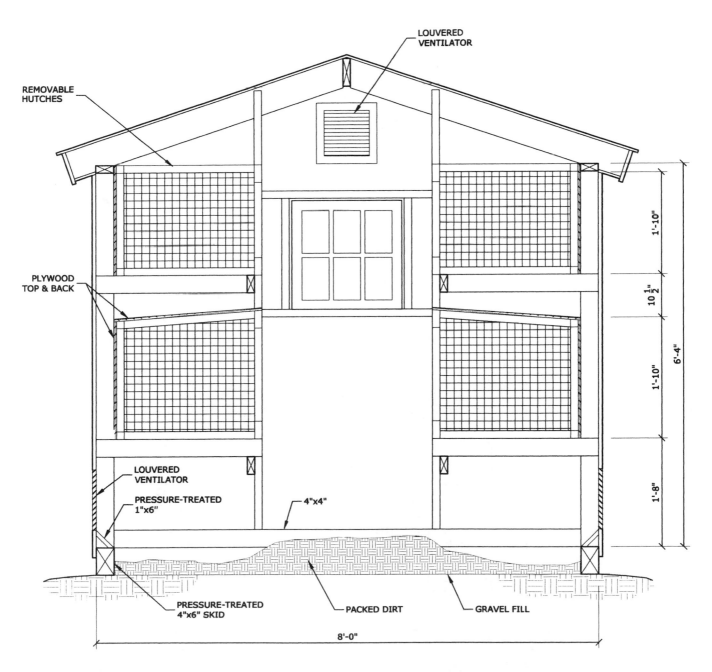

LOUVERED
VENTILATOR

REMOVABLE
HUTCHES

PLYWOOD
TOP & BACK

LOUVERED
VENTILATOR

PRESSURE-TREATED
1"x6"

4"x4"

PRESSURE-TREATED
4"x6" SKID

PACKED DIRT

GRAVEL FILL

1'-10"

10 1/2"

1'-10"

1'-8"

6'-4"

8'-0"

Section: double-deck wire-floor hutch

Removable Hutches

This design is for the removable hutches used in the rabbit house on page 42; they can also be hung in a barn, garage, or shed. If you don't want to build your own hutch from wood and wire, consider purchasing a pre-made all-wire cage or a kit. (See page 243 for plan credit.)

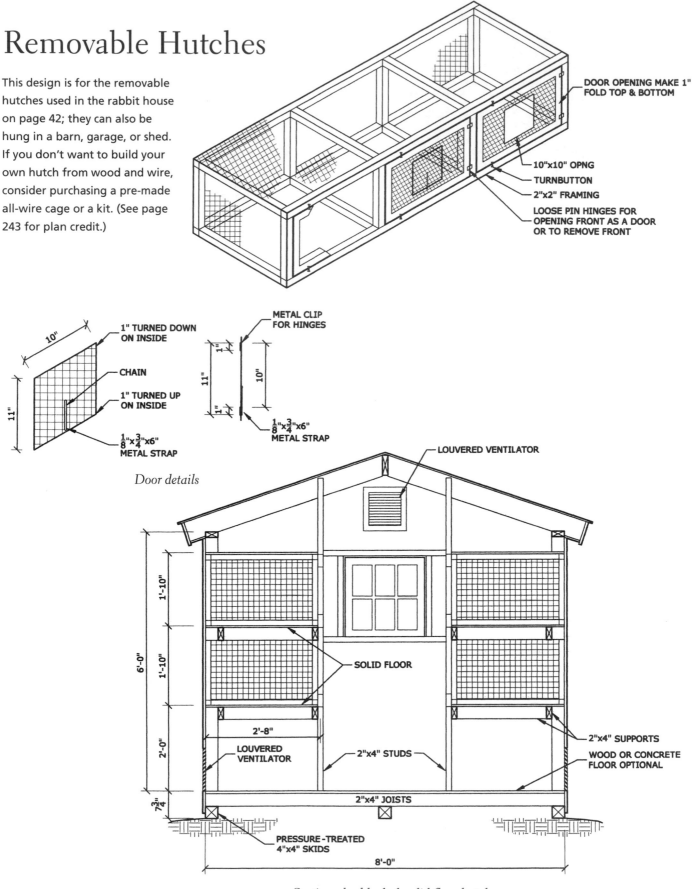

DOOR OPENING MAKE 1" FOLD TOP & BOTTOM

10"x10" OPNG

TURNBUTTON

2"x2" FRAMING

LOOSE PIN HINGES FOR OPENING FRONT AS A DOOR OR TO REMOVE FRONT

1" TURNED DOWN ON INSIDE

CHAIN

1" TURNED UP ON INSIDE

$\frac{1}{8}$"x$\frac{3}{4}$"x6" METAL STRAP

METAL CLIP FOR HINGES

$\frac{1}{8}$"x$\frac{3}{4}$"x6" METAL STRAP

Door details

LOUVERED VENTILATOR

1'-10"

6'-0"

1'-10"

2'-0"

7$\frac{3}{4}$"

SOLID FLOOR

2'-8"

LOUVERED VENTILATOR

2"x4" STUDS

2"x4" JOISTS

PRESSURE-TREATED 4"x4" SKIDS

8'-0"

2"x4" SUPPORTS

WOOD OR CONCRETE FLOOR OPTIONAL

Section: double-deck solid-floor hutch

MULTISPECIES SHELTERS

Sheep and goats, pigs, and calves all benefit from having access to some kind of small, portable shelter while on pasture. Several structures, such as Quonset-style, Port-A-Huts, and polymer calf domes, are available commercially, and there are a number of options for homemade structures.

With almost unlimited styles and options, portable structures are easy to construct and use. They can be covered in wood, farmer's tin, and even canvas. They may be open-ended, closed-ended with a door, or open-ended but with a tarp cover for shutting off the end in the worst weather.

Backyard pens work well for those who want to raise one or two calves, hogs, lambs, or kids during the summer. Backyard pens combine a shelter and a small paddock in the same structure, and they can be moved to fresh ground.

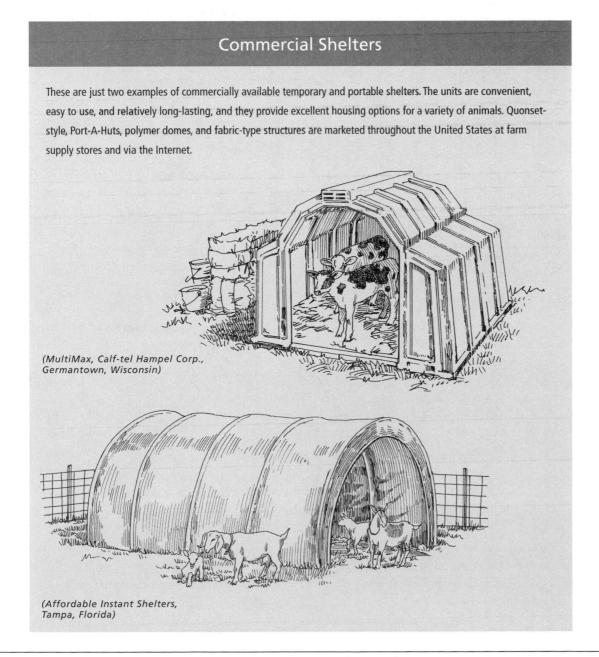

Commercial Shelters

These are just two examples of commercially available temporary and portable shelters. The units are convenient, easy to use, and relatively long-lasting, and they provide excellent housing options for a variety of animals. Quonset-style, Port-A-Huts, polymer domes, and fabric-type structures are marketed throughout the United States at farm supply stores and via the Internet.

(MultiMax, Calf-tel Hampel Corp., Germantown, Wisconsin)

(Affordable Instant Shelters, Tampa, Florida)

A-Frame Hut

A-frame huts are great for pasture-raised animals. Sows can farrow inside (the guard rail protects piglets), and sheep and goats use them in extreme weather. Place a door at one end only, leaving the other end open so animals can come and go at will. For sanitation, I prefer the unfloored version. (See page 243 for plan credit.)

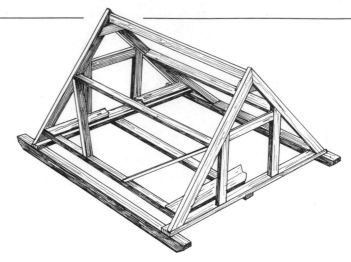

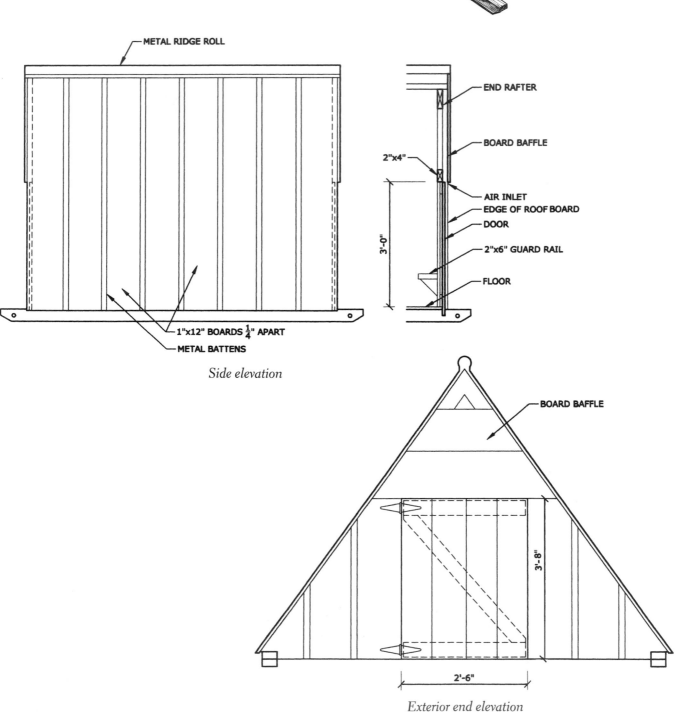

METAL RIDGE ROLL

1"x12" BOARDS ¼" APART

METAL BATTENS

Side elevation

END RAFTER

BOARD BAFFLE

2"x4"

AIR INLET

EDGE OF ROOF BOARD

DOOR

2"x6" GUARD RAIL

FLOOR

3'-0"

BOARD BAFFLE

3'-8"

2'-6"

Exterior end elevation

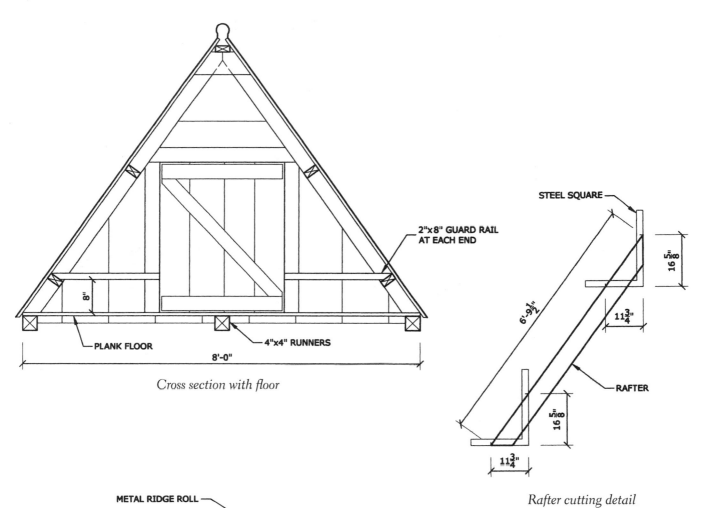

2"x8" GUARD RAIL AT EACH END

8"

PLANK FLOOR

4"x4" RUNNERS

8'-0"

Cross section with floor

STEEL SQUARE

$16\frac{5}{8}$"

6'-9$\frac{1}{2}$"

$11\frac{3}{4}$"

RAFTER

$16\frac{5}{8}$"

$11\frac{3}{4}$"

Rafter cutting detail

METAL RIDGE ROLL

2"x4" RIDGE

1"x6" CLEAT

2"x4" RAFTERS

1"x12" BOARDS

2"x4" GIRT

2"x8" GUARD RAIL AT EACH END

5'-6$\frac{1}{2}$"

3'-11"

8'-0"

Cross section without floor

Modified A-Frame Hut 1

This design is a little more spacious and works well for animals raised in a small pen. For ventilation, the interior wall extends only to the height of the collar beam, then boards are secured to the outside edge of the roof down to the level of the collar beam. (See page 243 for plan credit.)

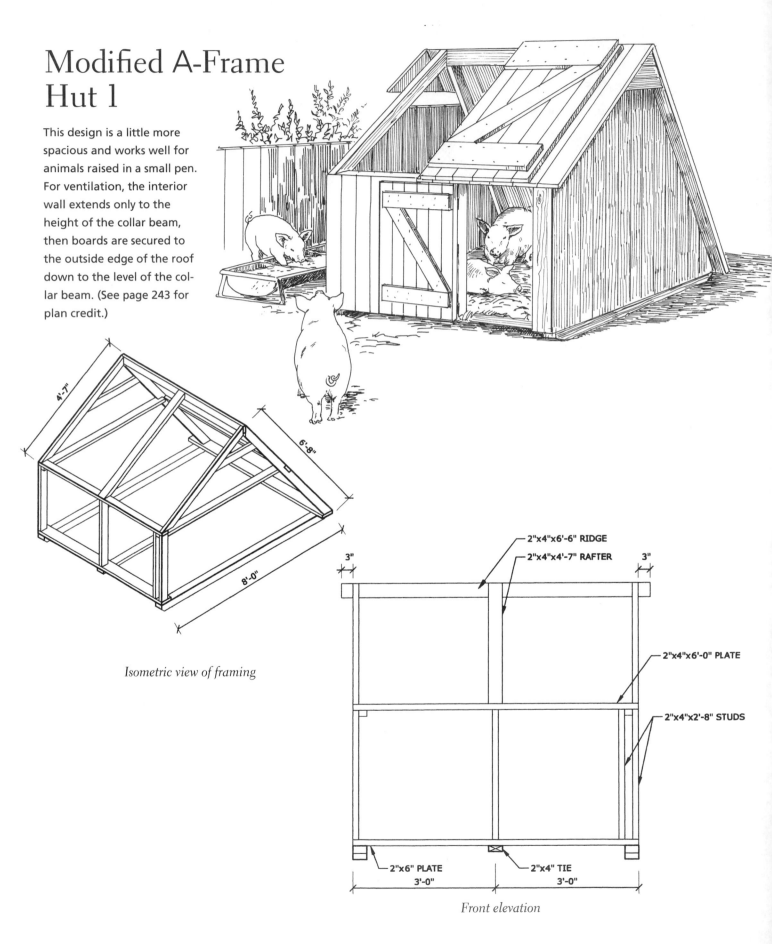

Isometric view of framing

4'-7"

6'-8"

8'-0"

3"

2"x4"x6'-6" RIDGE

2"x4"x4'-7" RAFTER

3"

2"x4"x6'-0" PLATE

2"x4"x2'-8" STUDS

2"x6" PLATE

2"x4" TIE

3'-0"

3'-0"

Front elevation

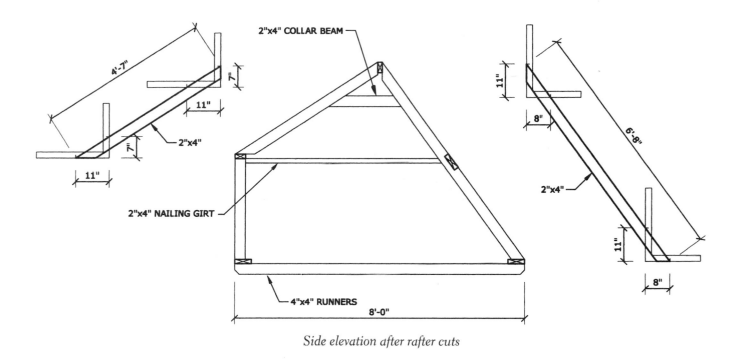

2"x4" COLLAR BEAM

4'-7"

7"

11"

7"

11"

2"x4"

2"x4" NAILING GIRT

4"x4" RUNNERS

8'-0"

11"

8"

6'-8"

2"x4"

11"

8"

Side elevation after rafter cuts

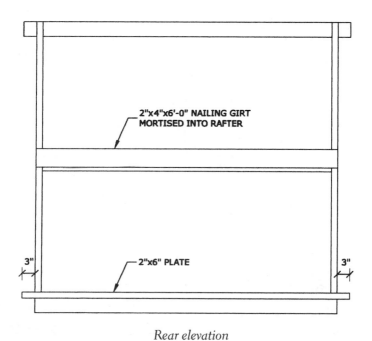

2"x4"x6'-0" NAILING GIRT
MORTISED INTO RAFTER

3"

2"x6" PLATE

3"

Rear elevation

Modified A-Frame Hut 2

Just slightly smaller than the previous plan, this version is a good substitute on pasture. The angle of the back wall allows piglets to move safely out of mom's way when she rolls around. (See page 243 for plan credit.)

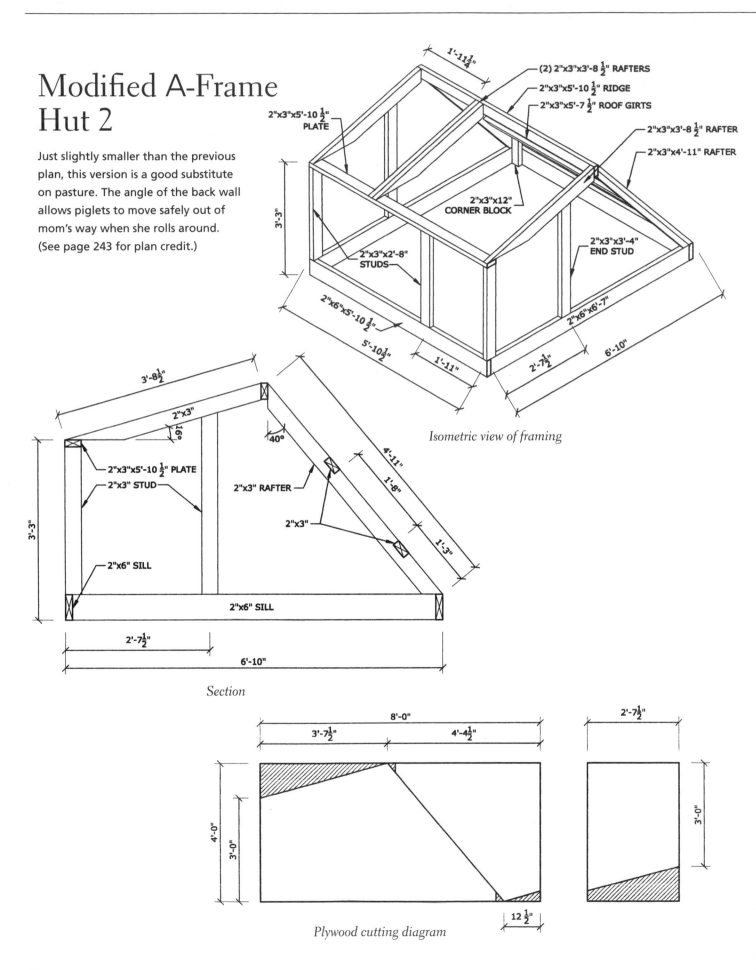

2"x3"x5'-10 ½" PLATE

1'-11 ¼"

(2) 2"x3"x3'-8 ½" RAFTERS

2"x3"x5'-10 ½" RIDGE

2"x3"x5'-7 ½" ROOF GIRTS

2"x3"x3'-8 ½" RAFTER

2"x3"x4'-11" RAFTER

2"x3"x12" CORNER BLOCK

2"x3"x3'-4" END STUD

3'-3"

2"x3"x2'-8" STUDS

2"x6"x5'-10 ½"

5'-10 ½"

1'-11"

2'-7 ½"

2"x6"x6'-7"

6'-10"

Isometric view of framing

3'-8 ½"

2"x3"

16°

40°

4'-11"

2"x3"x5'-10 ½" PLATE

2"x3" STUD

2"x3" RAFTER

2"x3"

1'-8"

1'-3"

3'-3"

2"x6" SILL

2"x6" SILL

2'-7 ½"

6'-10"

Section

8'-0"

3'-7 ½"

4'-4 ½"

2'-7 ½"

4'-0"

3'-0"

3'-0"

12 ½"

Plywood cutting diagram

Portable Shelter 1

This shelter is excellent for a variety of pasture-raised animals or for raising a few animals in a pen. Thanks to its height, larger animals like cows, llamas, and alpacas will use it. Tow the shelter to clean ground with a truck or tractor. (See page 244 for plan credit.)

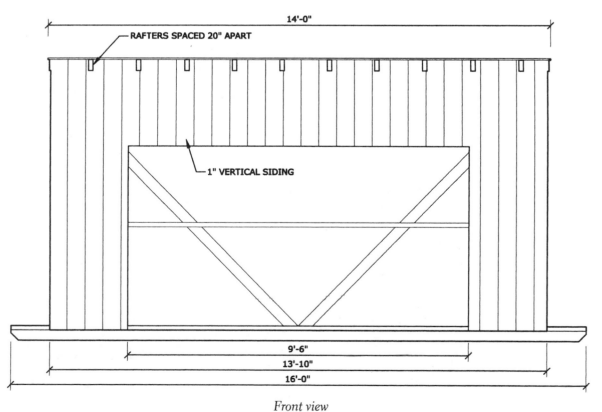

14'-0"

RAFTERS SPACED 20" APART

1" VERTICAL SIDING

9'-6"

13'-10"

16'-0"

Front view

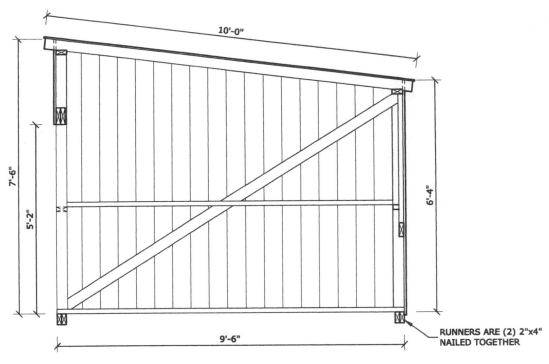

RUNNERS ARE (2) 2"x4"
NAILED TOGETHER

Cross-sectional view

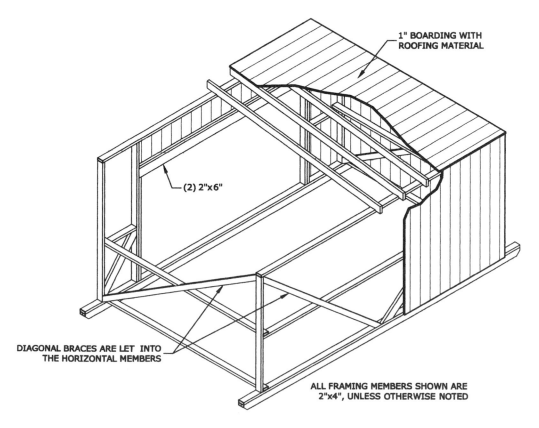

1" BOARDING WITH
ROOFING MATERIAL

(2) 2"x6"

DIAGONAL BRACES ARE LET INTO
THE HORIZONTAL MEMBERS

ALL FRAMING MEMBERS SHOWN ARE
2"x4", UNLESS OTHERWISE NOTED

Framing perspective

Portable Shelter 2

Traditionally, farms kept a sow or two, and this design was popular. The sow was kept year-round in a small pen with access to the shelter and was closed inside the shelter for farrowing. This structure is also suitable for a couple of sheep or goats kept in a small pen. With its swing-out side, it's easy to ventilate in hot weather. (See page 244 for plan credit.)

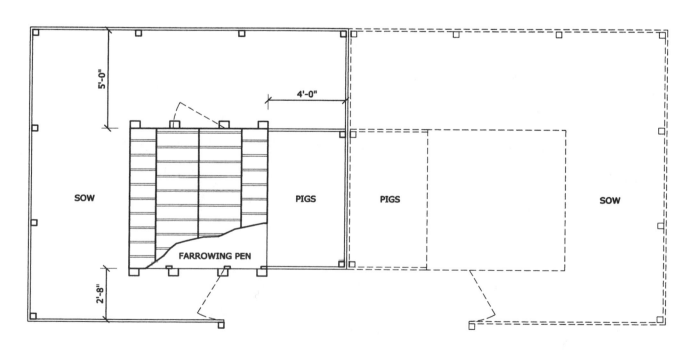

Plan showing single or double layout

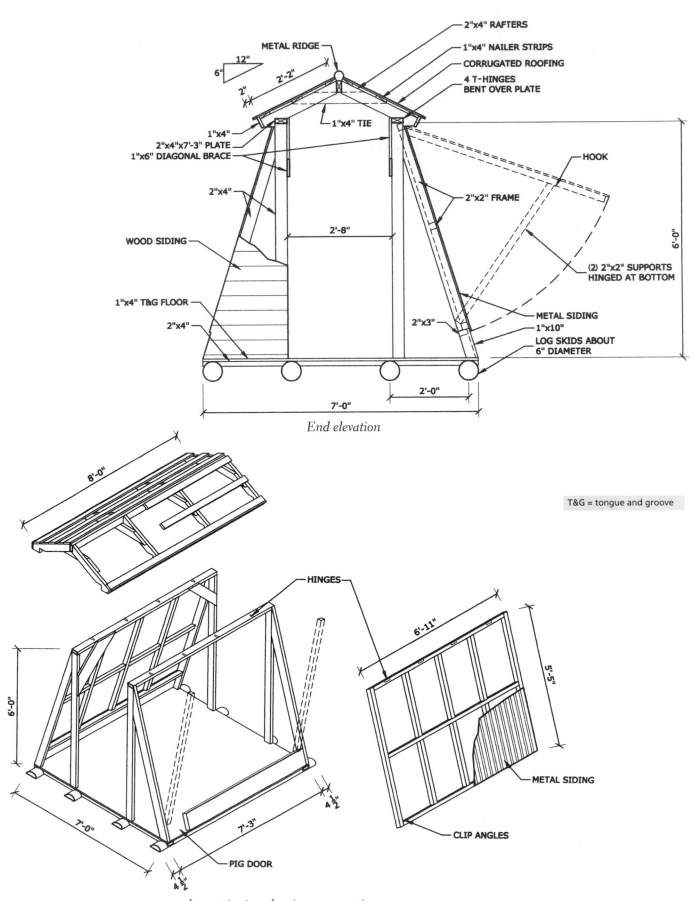

2"x4" RAFTERS
1"x4" NAILER STRIPS
CORRUGATED ROOFING
4 T-HINGES
BENT OVER PLATE

METAL RIDGE

12"
6"
2'-2"
2"

1"x4" TIE

1"x4"
2"x4"x7'-3" PLATE
1"x6" DIAGONAL BRACE

HOOK

2"x4"

2'-8"

2"x2" FRAME

WOOD SIDING

(2) 2"x2" SUPPORTS
HINGED AT BOTTOM

METAL SIDING

1"x4" T&G FLOOR

2"x4"

2"x3"
1"x10"
LOG SKIDS ABOUT
6" DIAMETER

6'-0"

2'-0"

7'-0"

End elevation

T&G = tongue and groove

HINGES

6'-11"

5'-5"

6'-0"

METAL SIDING

4 1/2"

7'-0"

7'-3"

CLIP ANGLES

PIG DOOR

4 1/2"

Isometric view showing construction

Backyard Pen 1

Portable backyard pens are great. If placed in the shade, you can easily raise a 4-H pig, lamb, or kid in one. Drill a few drainage holes in the flooring and cover with straw for animal comfort. (See page 244 for plan credit.)

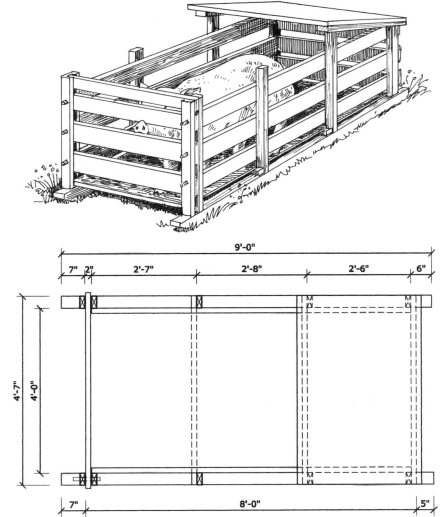

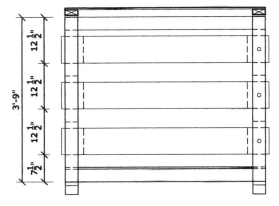

Elevation: gated end

CDX = construction-grade
Ø = diameter
PT = pressure-treated

Floor plan

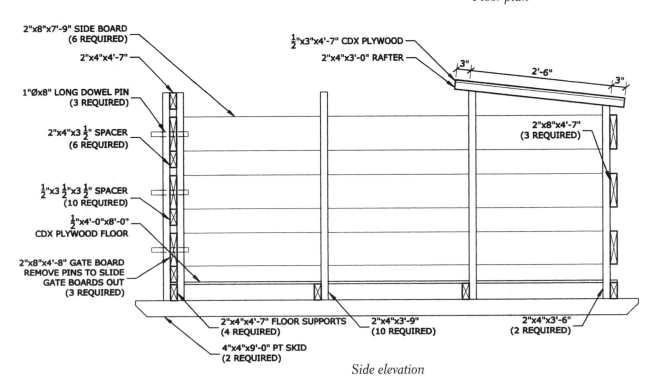

2"x8"x7'-9" SIDE BOARD
(6 REQUIRED)

2"x4"x4'-7"

1"Øx8" LONG DOWEL PIN
(3 REQUIRED)

2"x4"x3 ½" SPACER
(6 REQUIRED)

½"x3 ½"x3 ½" SPACER
(10 REQUIRED)

½"x4'-0"x8'-0"
CDX PLYWOOD FLOOR

2"x8"x4'-8" GATE BOARD
REMOVE PINS TO SLIDE
GATE BOARDS OUT
(3 REQUIRED)

½"x3"x4'-7" CDX PLYWOOD

2"x4"x3'-0" RAFTER

2"x8"x4'-7"
(3 REQUIRED)

2"x4"x4'-7" FLOOR SUPPORTS
(4 REQUIRED)

2"x4"x3'-9"
(10 REQUIRED)

2"x4"x3'-6"
(2 REQUIRED)

4"x4"x9'-0" PT SKID
(2 REQUIRED)

Side elevation

Calf Hutch with Movable Paddock

Raising a bottle calf can be challenging, but using a hutch and movable paddock is the easiest way to do it. The calf gets fresh air and sunshine as well as protection from rain, wind, and snow. These units also work well for a couple of sheep or goats. (See page 244 for plan credit.)

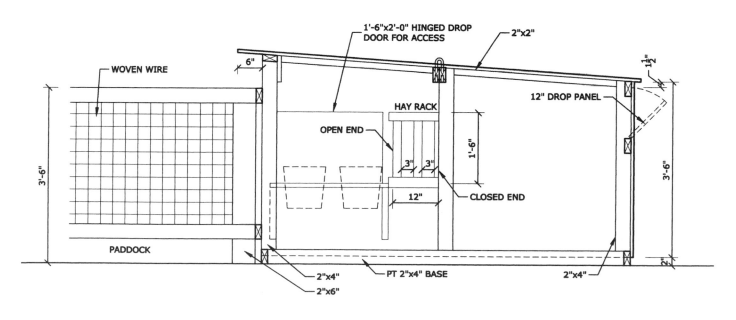

Side view

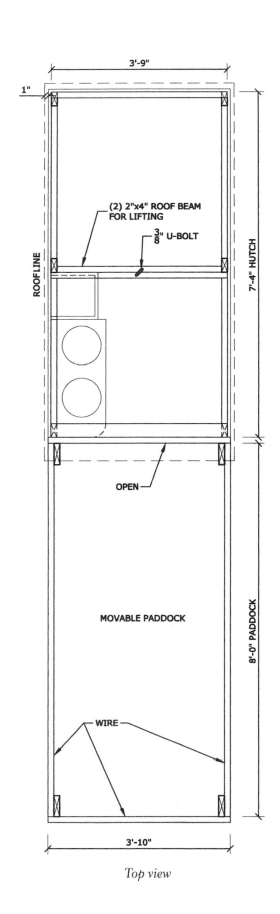

3'-9"

1"

ROOFLINE

(2) 2"x4" ROOF BEAM
FOR LIFTING

$\frac{3}{8}$" U-BOLT

7'-4" HUTCH

OPEN

MOVABLE PADDOCK

8'-0" PADDOCK

WIRE

3'-10"

Top view

PT = pressure-treated

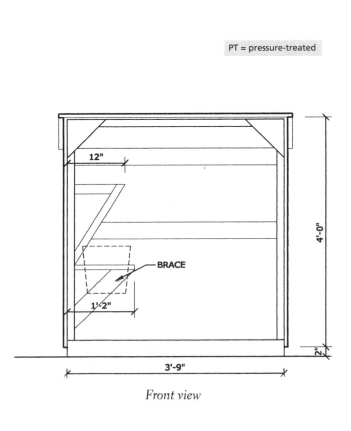

12"

BRACE

1'-2"

4'-0"

3'-9"

2"

Front view

BIRDS

Poultry and fowl are great additions to any farm, but they also do well as backyard animals. Many towns allow a small flock of hens, ducks, or geese as "pets." Six hens can produce most of the eggs your family needs, and fresh eggs are far superior to those available in stores. If you keep a rooster and hens from a heritage, dual-purpose breed known for meat and egg production and for hens that go "broody" and raise their own chicks, you can stay fully supplied with meat and eggs with little effort.

Ken and I are partial to Plymouth Barred Rocks and Jersey Giants. For a small flock of chickens (say up to a couple of dozen hens confined full time or fifty that are allowed out during the day), a 10-foot by 12-foot poultry shed on skids works very well. To provide daytime access to the outdoors, you can fence in a small area around the shed with portable electric netting, or, if neighbors and predators aren't a major concern, you can allow the birds to range freely. This style shed can also be used for other poultry or fowl.

An ark is an alternative for daytime hours in areas where predators or neighbors are a concern, or for flocks of fewer than a dozen birds. (There is a great Web site maintained by Katy Skinner called the City Chicken that has photos of lots of different arks. Folks simply share photos of their arks, and it's a great place to get ideas. For details, see the listing in the resources.) The ark can be moved each day, allowing a small flock to graze, eat bugs, and enjoy the sunshine while offering control and protection. An ark is also an excellent option for a couple of ducks or geese. Arks are available commercially from several suppliers.

Joel Salatin, Andy Lee, and Herman Beck-Chenoweth are among the pioneers of larger-scale pastured-poultry production using portable pens. These pens are economical and produce healthy, tasty birds with high consumer demand. Larger models are available commercially, or they can be built at home.

Heritage Breeds

Heritage breeds were once common in the United States but have fallen out of favor with industrial producers. The American Livestock Breeds Conservancy (ALBC) is a nonprofit organization dedicated to the conservation and promotion of more than one hundred breeds of poultry and fowl, as well as dozens of breeds of pigs, asses, horses, cows, sheep, and goats that are threatened with extinction. We support ALBC and encourage others to help preserve these animals. (See the resources at the back of the book for contact information.)

Poultry Shed

This handy, economical little coop is easy to build. It can be built on skids or on piers (as shown), and pieces of fiberglass can be used for windows (as shown). If you have access to used windows with wooden sashes, you can substitute them. One nice feature of the design is storage for feed and supplies. (See page 244 for plan credit.)

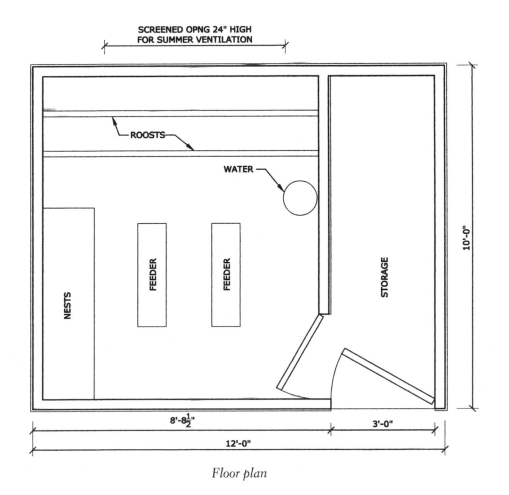

Floor plan

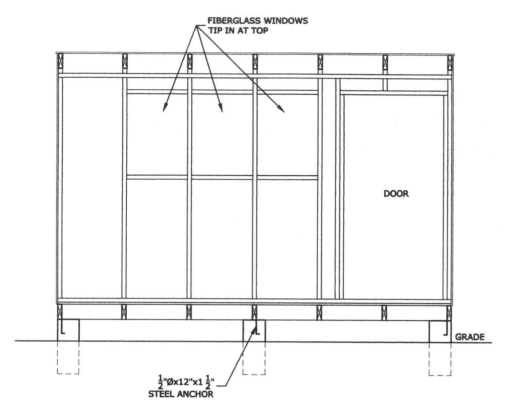

FIBERGLASS WINDOWS
TIP IN AT TOP

DOOR

GRADE

$\frac{1}{2}"Ø x 12" x 1\frac{1}{2}"$
STEEL ANCHOR

Front elevation framing

Ø = diameter
O.C. = on center
OPNG = opening

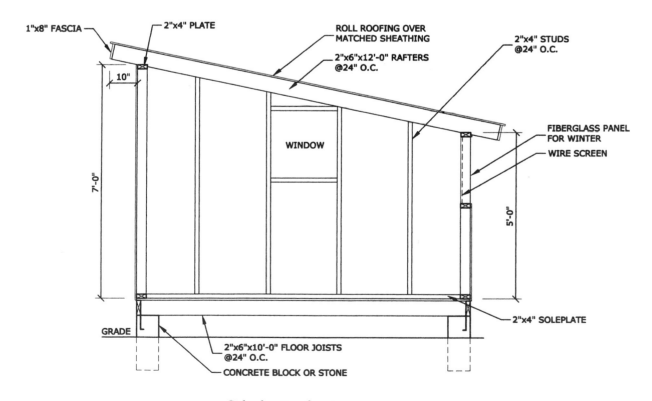

1"x8" FASCIA

2"x4" PLATE

ROLL ROOFING OVER
MATCHED SHEATHING

2"x6"x12'-0" RAFTERS
@24" O.C.

2"x4" STUDS
@24" O.C.

10"

WINDOW

FIBERGLASS PANEL
FOR WINTER

WIRE SCREEN

7'-0"

5'-0"

2"x4" SOLEPLATE

GRADE

2"x6"x10'-0" FLOOR JOISTS
@24" O.C.

CONCRETE BLOCK OR STONE

Side elevation framing

Portable Brooder House

This house, with its additional floor space for the chickens, works well as a brooder facility. The *hover* is the area where chicks are kept warm, usually with an electric heat lamp, until they feather out. Sections are provided for different window configurations. (See page 244 for plan credit.)

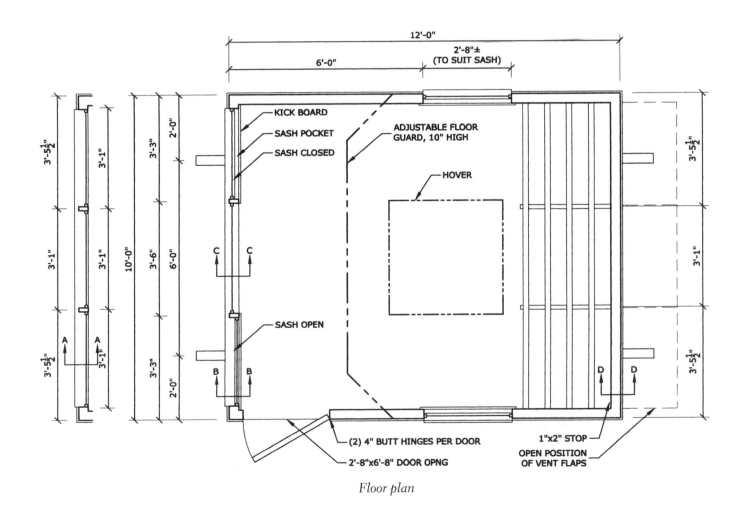

Floor plan

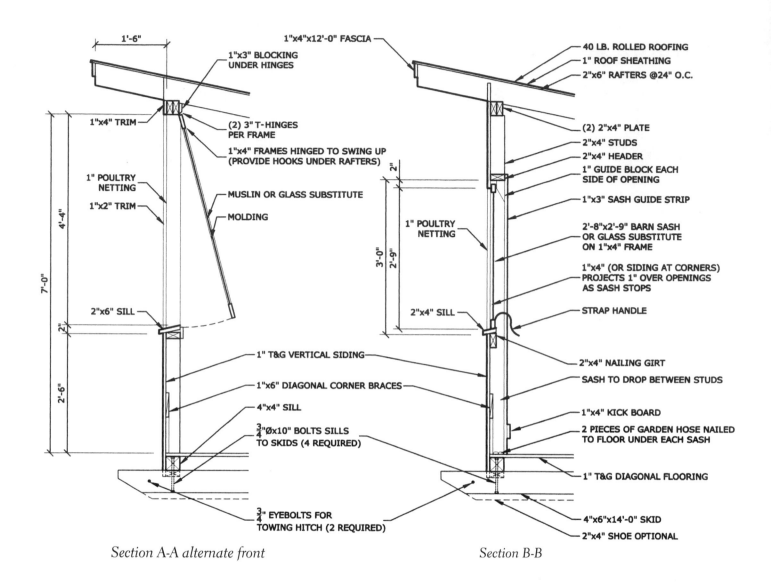

1'-6"

1"x4"x12'-0" FASCIA

1"x3" BLOCKING
UNDER HINGES

40 LB. ROLLED ROOFING
1" ROOF SHEATHING
2"x6" RAFTERS @24" O.C.

1"x4" TRIM

(2) 3" T-HINGES
PER FRAME

1"x4" FRAMES HINGED TO SWING UP
(PROVIDE HOOKS UNDER RAFTERS)

(2) 2"x4" PLATE
2"x4" STUDS
2"x4" HEADER

1" GUIDE BLOCK EACH
SIDE OF OPENING

1" POULTRY
NETTING

1"x2" TRIM

MUSLIN OR GLASS SUBSTITUTE

MOLDING

1"x3" SASH GUIDE STRIP

4'-4"

1" POULTRY
NETTING

2'-8"x2'-9" BARN SASH
OR GLASS SUBSTITUTE
ON 1"x4" FRAME

7'-0"

2"

3'-0"

2'-9"

1"x4" (OR SIDING AT CORNERS)
PROJECTS 1" OVER OPENINGS
AS SASH STOPS

2"x6" SILL

2"

2"x4" SILL

STRAP HANDLE

2'-6"

1" T&G VERTICAL SIDING

1"x6" DIAGONAL CORNER BRACES

4"x4" SILL

2"x4" NAILING GIRT

SASH TO DROP BETWEEN STUDS

1"x4" KICK BOARD

¾"Øx10" BOLTS SILLS
TO SKIDS (4 REQUIRED)

2 PIECES OF GARDEN HOSE NAILED
TO FLOOR UNDER EACH SASH

1" T&G DIAGONAL FLOORING

¾" EYEBOLTS FOR
TOWING HITCH (2 REQUIRED)

4"x6"x14'-0" SKID

2"x4" SHOE OPTIONAL

Section A-A alternate front

Section B-B

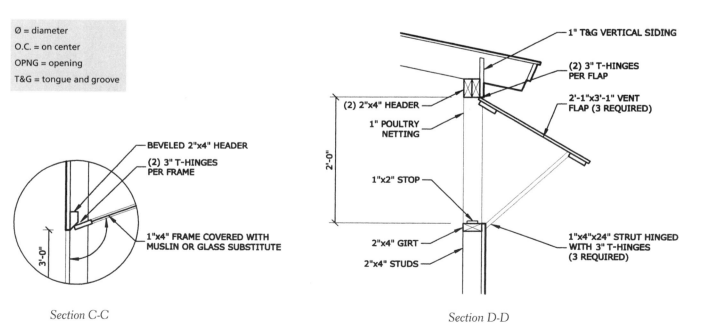

Ø = diameter
O.C. = on center
OPNG = opening
T&G = tongue and groove

1" T&G VERTICAL SIDING

(2) 3" T-HINGES
PER FLAP

2'-1"x3'-1" VENT
FLAP (3 REQUIRED)

(2) 2"x4" HEADER

1" POULTRY
NETTING

2'-0"

BEVELED 2"x4" HEADER

(2) 3" T-HINGES
PER FRAME

1"x2" STOP

3'-0"

2"x4" GIRT

2"x4" STUDS

1"x4" FRAME COVERED WITH
MUSLIN OR GLASS SUBSTITUTE

1"x4"x24" STRUT HINGED
WITH 3" T-HINGES
(3 REQUIRED)

Section C-C

Section D-D

Arks

When it comes to constructing arks, you are limited only by your imagination. These easy-to-build models are suitable for poultry and fowl. They can be used during the day for grazing or serve as full-time housing if protected from predators by an electric fence.

This ark readily accommodates a small flock. Because the nest box is mounted on the side and has a roof that flips open, it's easy to gather eggs.

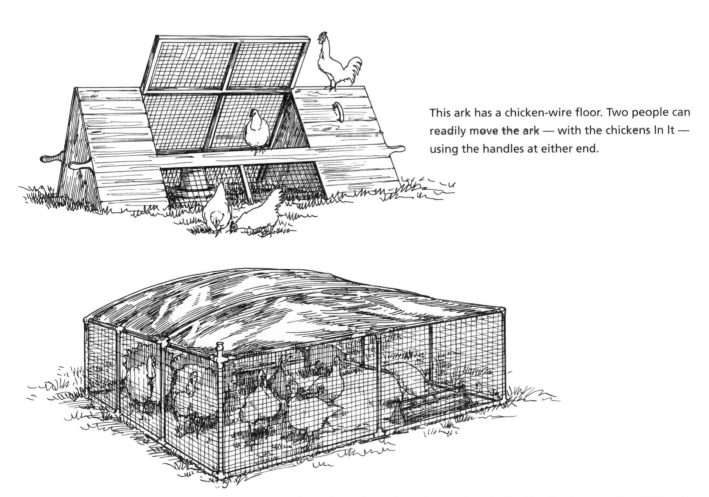

This ark has a chicken-wire floor. Two people can readily move the ark — with the chickens in it — using the handles at either end.

This is the easiest of the three arks to build. Simply glue together PVC and cover with chicken wire. This design works well for larger numbers of birds and is used by many commercial "pasture poultry" producers.

Summer Range Shelter

This traditional design for a summer pasture shelter predates arks. It is a little larger and more complicated to build but will last for years. Leave out the roosts, and it can work effectively for turkeys. (See page 244 for plan credit.)

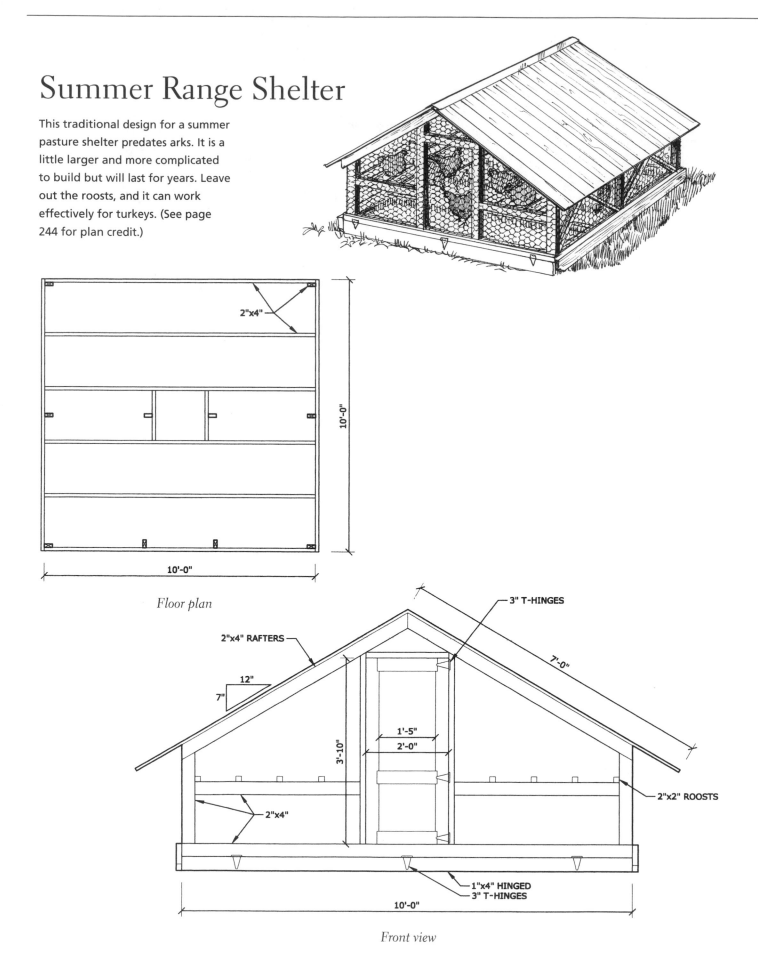

2"x4"

10'-0"

10'-0"

Floor plan

3" T-HINGES

2"x4" RAFTERS

7'-0"

12"

7"

3'-10"

1'-5"

2'-0"

2"x4"

2"x2" ROOSTS

1"x4" HINGED
3" T-HINGES

10'-0"

Front view

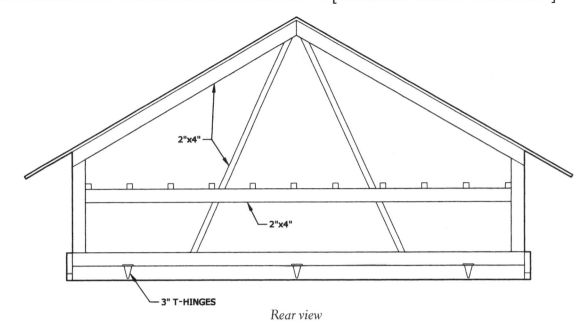

2"x4"

2"x4"

3" T-HINGES

Rear view

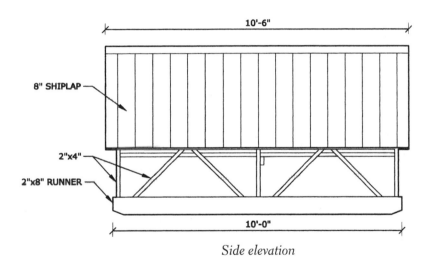

10'-6"

8" SHIPLAP

2"x4"

2"x8" RUNNER

10'-0"

Side elevation

O.C. = on center

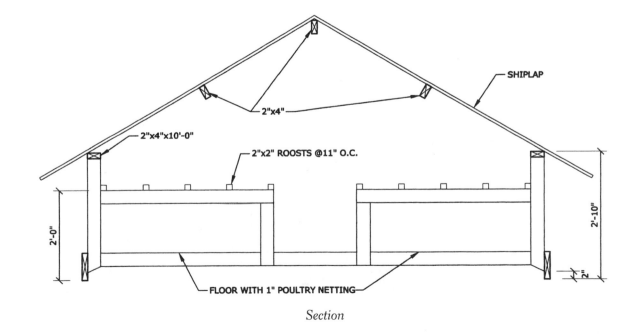

SHIPLAP

2"x4"

2"x4"x10'-0"

2"x2" ROOSTS @11" O.C.

2'-0"

2'-10"

2"

FLOOR WITH 1" POULTRY NETTING

Section

SKID BARNS

Small barns on skids are suitable for a variety of livestock species. These units can be moved easily with a tractor or pickup; they typically have a shed roof or gable roof. The portable stable is excellent for a single horse or cow and is constructed without a wooden floor. The shed, with a wood floor, works well for a few pigs, sheep, or goats (or llamas or alpacas).

Portable Stable

If you have limited construction experience and need animal housing, this is a good project to learn on. It's modestly challenging but doable. It's also practical to have around. (See page 244 for plan credit.)

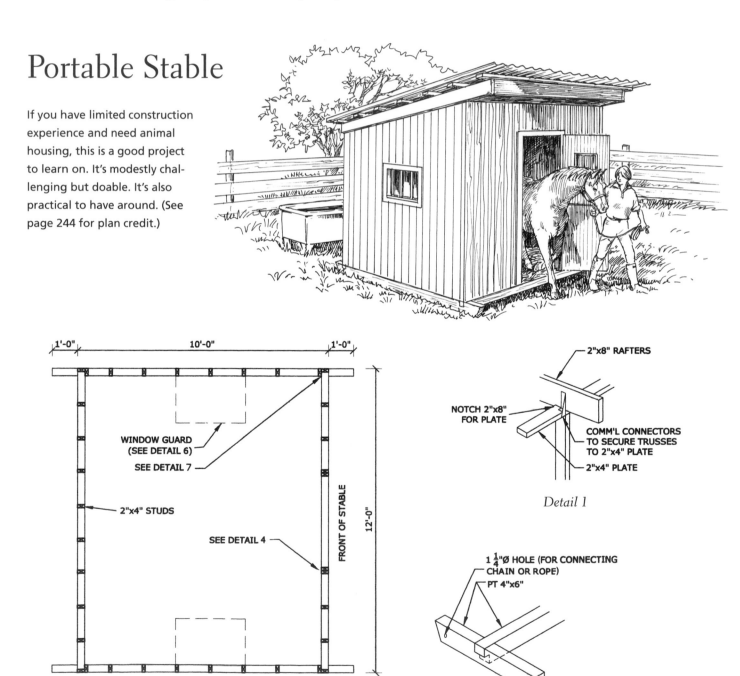

Floor plan

Detail 1

Detail 2

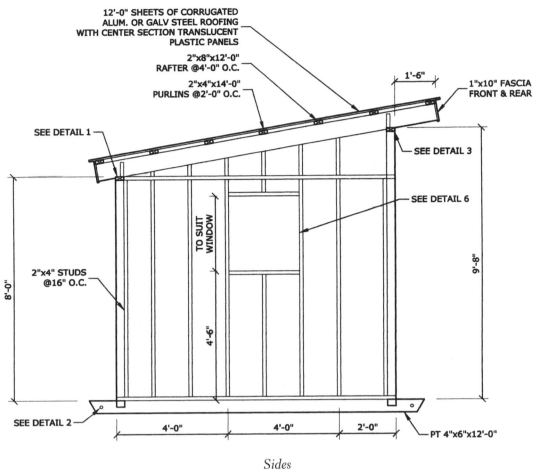

12'-0" SHEETS OF CORRUGATED
ALUM. OR GALV STEEL ROOFING
WITH CENTER SECTION TRANSLUCENT
PLASTIC PANELS

2"x8"x12'-0"
RAFTER @4'-0" O.C.

2"x4"x14'-0"
PURLINS @2'-0" O.C.

1'-6"

1"x10" FASCIA
FRONT & REAR

SEE DETAIL 1

SEE DETAIL 3

SEE DETAIL 6

TO SUIT WINDOW

2"x4" STUDS
@16" O.C.

8'-0"

9'-8"

4'-6"

SEE DETAIL 2

4'-0"   4'-0"   2'-0"

PT 4"x6"x12'-0"

Sides

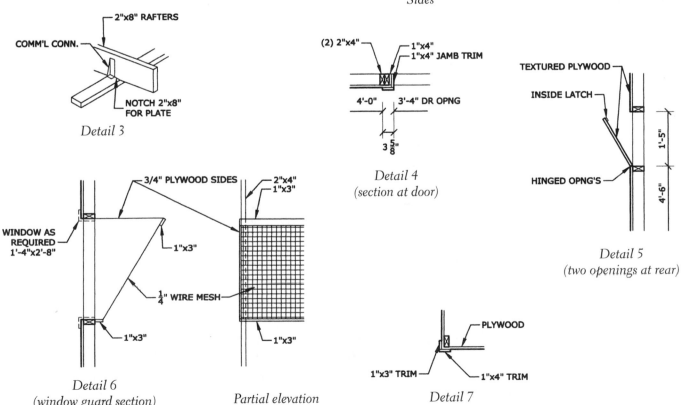

2"x8" RAFTERS

COMM'L CONN.

NOTCH 2"x8"
FOR PLATE

Detail 3

(2) 2"x4"

1"x4"

1"x4" JAMB TRIM

4'-0"   3'-4" DR OPNG

3 5/8"

Detail 4
(section at door)

TEXTURED PLYWOOD

INSIDE LATCH

HINGED OPNG'S

1'-5"

4'-6"

Detail 5
(two openings at rear)

3/4" PLYWOOD SIDES

2"x4"
1"x3"

WINDOW AS
REQUIRED
1'-4"x2'-8"

1"x3"

1/4" WIRE MESH

1"x3"

1"x3"

Detail 6
(window guard section)

Partial elevation

PLYWOOD

1"x3" TRIM

1"x4" TRIM

Detail 7

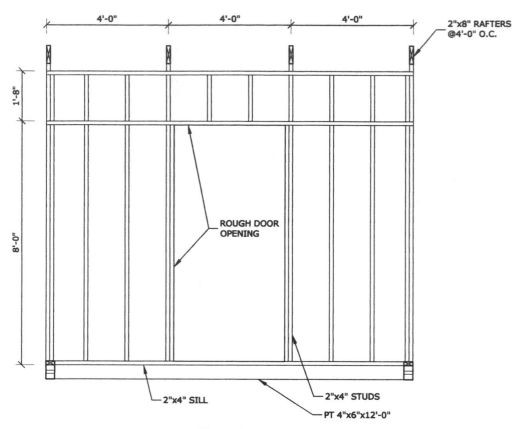

2"x8" RAFTERS @4'-0" O.C.

4'-0"   4'-0"   4'-0"

1'-8"

8'-0"

ROUGH DOOR OPENING

2"x4" SILL

2"x4" STUDS

PT 4"x6"x12'-0"

Front view

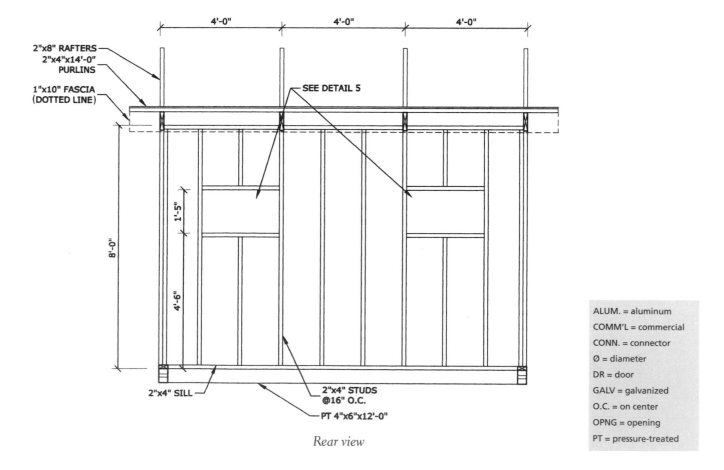

2"x8" RAFTERS
2"x4"x14'-0" PURLINS

1"x10" FASCIA (DOTTED LINE)

SEE DETAIL 5

4'-0"   4'-0"   4'-0"

8'-0"

1'-5"

4'-6"

2"x4" SILL

2"x4" STUDS @16" O.C.

PT 4"x6"x12'-0"

Rear view

ALUM. = aluminum
COMM'L = commercial
CONN. = connector
Ø = diameter
DR = door
GALV = galvanized
O.C. = on center
OPNG = opening
PT = pressure-treated

Movable Shed

This type of shed can be used for pigs, sheep, and goats. If sows and their litters will use it, include a guard rail so the piglets won't be rolled on. Drill a few drainage holes in the flooring, and cover it with straw to make the animals comfortable. (See page 244 for plan credits.)

ANCHOR STRAP

Perspective of framing

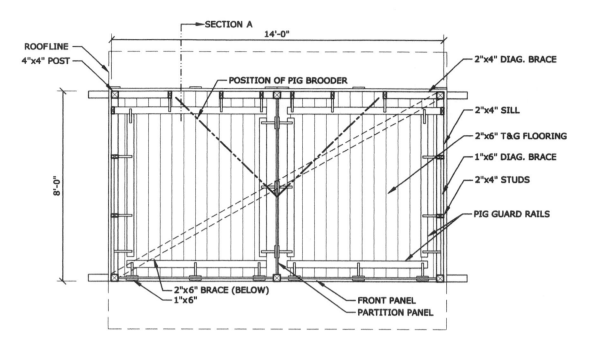

SECTION A

14'-0"

ROOFLINE
4"x4" POST

POSITION OF PIG BROODER

2"x4" DIAG. BRACE

2"x4" SILL

2"x6" T&G FLOORING

1"x6" DIAG. BRACE

2"x4" STUDS

PIG GUARD RAILS

8'-0"

2"x6" BRACE (BELOW)
1"x6"

FRONT PANEL
PARTITION PANEL

Floor plan

CAR. = carriage
CDX = construction-grade
DIAG. = diagonal
Ø = diameter
O.C. = on center
T&G = tongue and groove

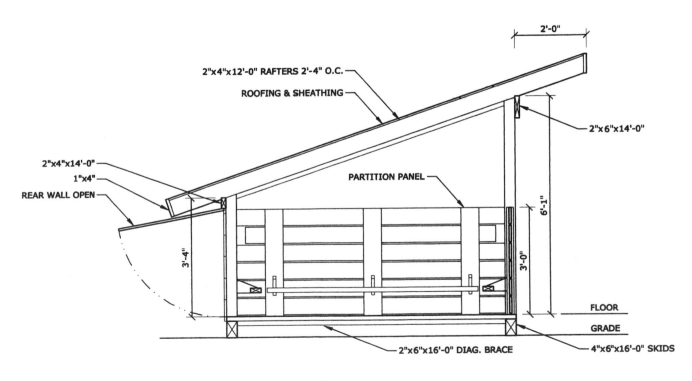

2'-0"

2"x4"x12'-0" RAFTERS 2'-4" O.C.

ROOFING & SHEATHING

2"x6"x14'-0"

2"x4"x14'-0"
1"x4"

REAR WALL OPEN

PARTITION PANEL

6'-1"

3'-4"

3'-0"

FLOOR

GRADE

2"x6"x16'-0" DIAG. BRACE

4"x6"x16'-0" SKIDS

Section A-A

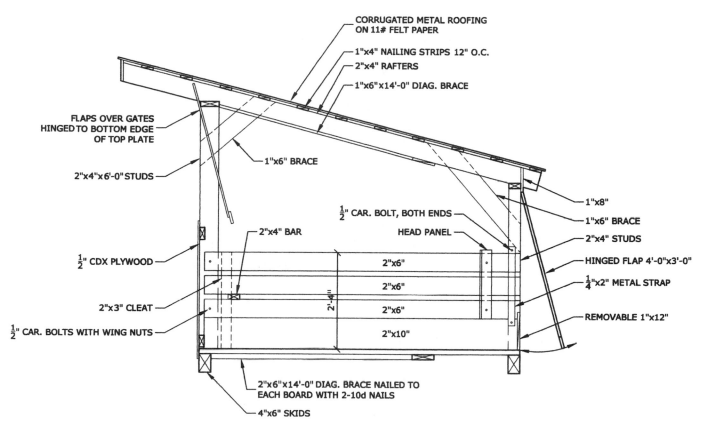

CORRUGATED METAL ROOFING
ON 11# FELT PAPER

1"x4" NAILING STRIPS 12" O.C.

2"x4" RAFTERS

1"x6"x14'-0" DIAG. BRACE

FLAPS OVER GATES
HINGED TO BOTTOM EDGE
OF TOP PLATE

1"x6" BRACE

2"x4"x6'-0" STUDS

½" CAR. BOLT, BOTH ENDS

2"x4" BAR

HEAD PANEL

1"x8"

1"x6" BRACE

2"x4" STUDS

½" CDX PLYWOOD

2"x6"

2"x6"

HINGED FLAP 4'-0"x3'-0"

2"x3" CLEAT

2"x6"

¼"x2" METAL STRAP

2'-4"

2"x6"

½" CAR. BOLTS WITH WING NUTS

2"x10"

REMOVABLE 1"x12"

2"x6"x14'-0" DIAG. BRACE NAILED TO
EACH BOARD WITH 2-10d NAILS

4"x6" SKIDS

Alternative approach (front and side panels are plywood instead of boards;
solid front can be let down in severe winter weather)

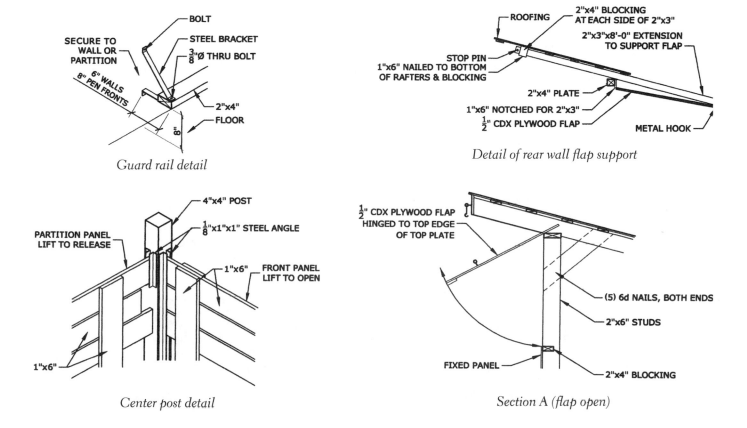

BOLT

SECURE TO
WALL OR
PARTITION

STEEL BRACKET

⅜"Ø THRU BOLT

8" 6" WALLS
PEN FRONTS

2"x4"

8"

FLOOR

Guard rail detail

2"x4" BLOCKING
AT EACH SIDE OF 2"x3"

ROOFING

2"x3"x8'-0" EXTENSION
TO SUPPORT FLAP

STOP PIN
1"x6" NAILED TO BOTTOM
OF RAFTERS & BLOCKING

2"x4" PLATE

1"x6" NOTCHED FOR 2"x3"

½" CDX PLYWOOD FLAP

METAL HOOK

Detail of rear wall flap support

4"x4" POST

⅛"x1"x1" STEEL ANGLE

PARTITION PANEL
LIFT TO RELEASE

1"x6"

FRONT PANEL
LIFT TO OPEN

1"x6"

Center post detail

½" CDX PLYWOOD FLAP
HINGED TO TOP EDGE
OF TOP PLATE

(5) 6d NAILS, BOTH ENDS

2"x6" STUDS

FIXED PANEL

2"x4" BLOCKING

Section A (flap open)

Janet McNally

Janet McNally, a shepherd from Hinckley, Minnesota, knows that portable structures can be *really portable*, and temporary. In the mid-nineties, Janet had already been lambing on pasture in spring as part of an intensively managed grazing operation. During dry weather, the ewes and lambs did wonderfully and needed no shelter at all, but spring in Minnesota can be cold and wet, so lambs born during these soupy periods often needed help. She would gather lambs that were beginning to show signs of hypothermia and take them and their mamas to the barn in a wagon. As Janet puts it, "It was not only time-consuming, but you're doing it during the worst weather. I realized I was fighting gravity."

"The thing that's neat about these tepees . . . they're easily expandable as you grow your operation."

Janet began looking for a way to bring the barn to her charges in the field: "I was kind of thinking about different ideas when, honestly, I saw a painting in an animal scientist's office. It showed a technique used by shepherds on the western range in times past: the landscape was dotted with tepees and sheep, and that's where I got the idea."

Janet contacted Jim Walker, a descendant of a long line of ranchers, who also runs Walker's Pack Saddlery, an Oregon-based company specializing in tents for horse camping and hunting. Jim was able to make some small modifications to his 6-foot by 6-foot "range tepee," a pyramidal-style tent that perfectly met Janet's need.

The tepees are not costly, and they last a decade or so. For her flock of 150 ewes, Janet gets by with just six tepees, making them a great deal when you consider how many lambs you put in them over time.

The lambs go into the tepees if their temperature falls to 101°F (if their temperature goes below that, they need more serious treatment), which is marginally cold for lambs, but they quickly recover once out of the weather. "I run around with a thermometer," Janet explains. "If things are getting dicey and I see lambs in that 101°F to 102°F temperature range, the whole family goes in. It's really only the newborns that are vulnerable, and it is almost always the ones that are born in the rain that have to go in. If they are born before the rain and their coat dries out, they seem to be able to handle the rain without intervention."

Janet reports that the animal's body heat quickly warms the tent, making a comfy shelter for the babies to dry off. As soon as they're dry, they regain the ability to modulate their own body temperature, so most of the time they can be let out within twenty-four hours. Occasionally, she also uses the tepees with a yearling ewe that doesn't want to "mother up." She reports that when the ewe and lamb(s) are placed in the tepee, with all distractions removed, the yearling generally bonds quickly with its offspring.

Although there's a full set of buildings on the farm Janet rents, she rarely uses them because they're not a good set of buildings — they are too low and muddy — so "there are more health problems in them than they are worth," she says.

"The thing that's neat about these tepees, and that's neat about portable fences, is that they're easily expandable as you grow your operation. If you have twenty sheep, you get what you need for them now; later, when you have a hundred, you get more. You grow your system in a logical order, as you need it!"

5. WINDBREAKS & SHADE SHELTERS

Livestock are warm-blooded, and warm-blooded creatures need to maintain their body temperature within a relatively narrow range. When air temperatures exceed this range or fall below it, animals expend extra energy. Under extreme conditions at either end of the temperature spectrum, animals perform poorly and are more vulnerable to disease and other health problems, including death. But windbreaks and shade shelters mitigate the extremes and minimize the health problems that often plague unprotected animals.

Take a beef cow, for example. A mature cow in good condition with a heavy winter coat is fine at temperatures down to 18°F (adjusted for windchill), but below that she needs 13 percent more energy for every ten-degree decline in windchill temperature. If her coat happens to be wet, the minimum acceptable base temperature climbs to 59°F. Mature animals in good condition have lower thresholds for critical temperature than do youngsters, so a temperature that suits a mature cow will kill her newborn calf and challenge a yearling. Pigs and poultry have higher thresholds for critical temperatures, while sheep, with their wonderful wool coats, have lower thresholds.

Wind speed	Temperature (°F)*											
	−10	−5	0	5	10	15	20	25	30	35	40	45
0	−10	−5	0	5	10	15	20	25	30	35	40	45
5	−22	−16	−11	−5	1	7	13	19	25	31	36	42
10	−28	−22	−16	−10	−4	3	9	15	21	27	34	40
15	−32	−26	−19	−13	−7	0	6	13	19	25	32	38
20	−35	−29	−22	−15	−9	−2	4	11	17	24	30	37
25	−37	−31	−24	−17	−11	−4	3	9	16	23	29	36
30	−39	−33	−26	−19	−12	−5	1	8	15	22	28	35
35	−41	−34	−27	−21	−14	−7	0	7	14	21	28	35
40	−43	−36	−29	−22	−15	−8	−1	6	13	20	27	34

WINDCHILL FACTORS FOR LIVESTOCK

Data from the National Weather Service.
*To convert degrees Fahrenheit to degrees Celsius, subtract 32 from the Fahrenheit temperature, then multiply the result by ⅝.

At the other extreme, temperatures above 75°F can begin to cause heat stress, and just as wind exacerbates low temperatures, humidity increases the problems associated with high temperatures. When temperatures get into the nineties, severe stress occurs, and temperatures in excess of 105°F can induce spontaneous heart failure. In one Midwest study, animals that were provided shade during the heat of summer gained as much as 22 percent more weight than did their counterparts without shade.

The National Weather Service, in cooperation with several land-grant universities, has developed a heat-stress index based on temperature and relative humidity. Again, pigs and poultry, as well as sheep in full fleece, are more likely to suffer as temperatures rise. Females in late pregnancy are more susceptible to problems, including abortion, than those in early pregnancy, and breeding males are especially susceptible to heat stress, with significant reduction in fertility as the first symptom. Not surprisingly, youngsters are more susceptible to heat stress than mature animals.

The best way to provide windbreaks and shade for animals is planting trees and shrubs, but animals — both domestic and wild — need protection today, and a natural windbreak might require decades to grow before it can provide optimum results. Planting trees and shrubs is an excellent undertaking, but while the plants grow, you can provide supplemental protection by building windbreak fences and shade shelters (see page 76ff).

NATURAL WINDBREAKS

The design of a natural windbreak depends in part on its use and the amount of snow in the area. For example, a *field windbreak* is less dense, with fewer rows of trees and shrubs, than one that will protect a house or barn, whereas one planted to protect a livestock pasture will have more evergreens. As a rule of thumb, the best layout for a windbreak is perpendicular, or at a right angle, to the prevailing wind. Luckily, windbreaks don't have to be in perfect, straight rows as is the case in the traditional farm shelterbelt. A more aesthetically pleasing appearance is easy to achieve by using a slightly random, curved pattern of trees and shrubs.

Whichever approach you use, if you plan to protect buildings and pasture where livestock will spend the winter, incorporate a mixture of trees and shrubs of varying sizes and a mix of evergreens and deciduous species. (If space limits the number of rows you can plant to one or two, choose evergreens over deciduous species.) This mix provides *porosity*, or air gaps, which reduces wind speed more effectively than a single type of plant. A windbreak can significantly reduce wind speed for a distance of fifteen times the height of the tallest trees, and it will provide some protection up to thirty times the height of the tallest trees.

Wider spacing of rows is effective in high snowfall areas, since more of the snow is captured within the windbreak than downwind of it, as would occur in tightly spaced rows. Spacing within rows varies by plant type, and it can start out closer than typically recommended, but as the windbreak matures, you will need to thin it to greater spacing to

HEAT-STRESS INDEX			
		Relative Humidity (%)	
Temperature (°F)*	Alert	Danger	Emergency
75	75	NA	NA
80	55	90	NA
85	30	60	90
90	15	35	65
95	0	20	45
100	0	10	30
105	0	0	5

Data from the National Weather Service.
*To convert degrees Fahrenheit to degrees Celsius, subtract 32 from the Fahrenheit temperature, then multiply the result by ⁵⁄₉.

Traditional shelterbelts are planted with straight rows of trees and shrubs; they significantly reduce wind and create snow deposition areas just beyond the shelterbelt.

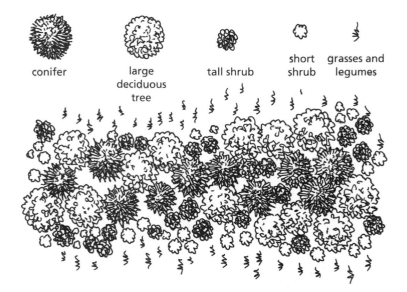

conifer   large deciduous tree   tall shrub   short shrub   grasses and legumes

On smaller parcels of land, a more natural appearance for a windbreak is easy to achieve with a slightly random pattern of trees and shrubs. As trees and shrubs mature, they will need to be thinned to provide air gaps.

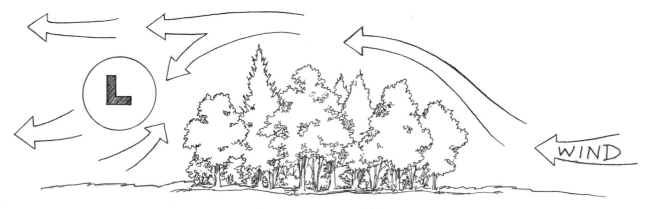

WIND

When wind encounters a natural or man-made windbreak, it rises up and over the top. On the lee side, turbulence is created by a low-pressure vacuum. Windbreak porosity (20 percent is the ideal) reduces the vacuum, resulting in less turbulence.

NATURAL WINDBREAKS			
Type	No. of Rows	No. of Rows Dense Conifer	Space between Rows in ft.
House/barn	4–10	2–4	12–20
Livestock pasture	4–10	3–6	12–20
Fields	1–2	1	10–15

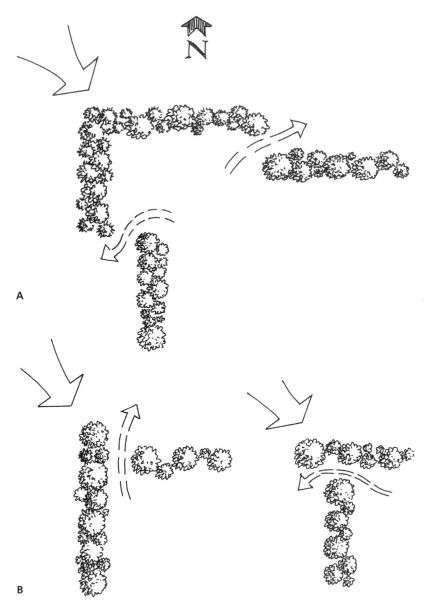

Gaps in windbreaks, for example, where lanes pass through the windbreak, create a funnel for the wind, which can result in snow being deposited in the middle of the lane. Avoid the negative effects of straight cuts by planting the windbreak to create offset gaps. Illustrations A and B show offset windbreak designs that reduce the negative impacts of a straight cut.

maintain its health. Carefully plan where you make openings in a shelterbelt, as some openings actually increase wind speed and snow deposit in sites where you least want it, like the road passing through a shelterbelt!

The best time of year to plant seedlings is in spring before the buds begin to swell. Summer and fall plantings are usually not as successful. Try to minimize stress on the seedlings during planting by working on a calm or cloudy day, and keep the trees moist. Avoid exposing seedlings, especially bare-root conifers, to the air any longer than necessary because it may damage their roots. Keep them in moist potting soil in a plastic bucket or the packaging they came in until they're ready to go in the ground, and keep them cool while they await planting.

After the windbreak is planted, it may require irrigation to get going, and it will need some weed and grass control. We have successfully planted windbreaks in grassed areas without resorting to herbicides to kill the grass, but it takes a commitment to diligent hand mowing for several years until the trees and shrubs are taller than the grass. Eventually the trees and shrubs will shade out most of the grass and weeds.

Livestock must be fenced out of the windbreak, especially in its early years, or they will trample and eat the plantings. When the windbreak is fully mature, it may benefit from short rotational grazing spurts by livestock, but even a mature break can sustain serious damage if animals are allowed routine access.

WINDBREAK FENCES

Windbreak fences offer protection in open areas until natural windbreaks become established. In Canada, where they know all about deep-freeze windchill factors, extensive research has been done on the best designs for windbreak fences, and it has been found that optimal protection is offered by a fence with about 80 percent solid surface area and

20 percent vertical openings of less than 4 inches wide. So, if using 1x10 boards, for example, you would space them 2.5 inches apart (10÷4 = 2.5). If using 1x8s, the spacing would be 2 inches (8÷4 = 2.0). This finding is perhaps counterintuitive: you might assume that a solid wall would provide better protection than a wall with openings, but the solid wall creates a violent and turbulent downdraft on the leeward side of the fence because of an air-pressure vacuum on that side. When small openings are left in the fence, there is no vacuum on the leeward side and hence no turbulent downdraft.

Windbreak fences can be constructed with boards, plywood, light logs (like aspen or birch), or corrugated metal. Meeting the criteria for ratio of solid to open surface and width of opening with wide sheets such as plywood or corrugated metal requires cutting the sheets into strips no more than 16 inches wide.

Like natural windbreaks, fences also provide good protection for about fifteen times the height of the fence. Thus, a 12-foot-high fence will protect an area 120 to 180 feet downwind. The majority of the snow will drop within roughly the first 25 percent of the protected area, so for the same 12-foot-high fence, the snow will drop most of its load within 45 feet of the fence. Fences are also most effective when placed at right angles to the prevailing winds. Portable fences are convenient when stock is moved from paddock to paddock during the winter or for dealing with odd wind patterns.

A porous fence is an excellent windbreak if it provides 20 percent vertical openings. *Inset:* Detail of backboard connections on a windbreak fence. Horizontal backboards are run between fence posts, and vertical fence boards are screwed into the backboards.

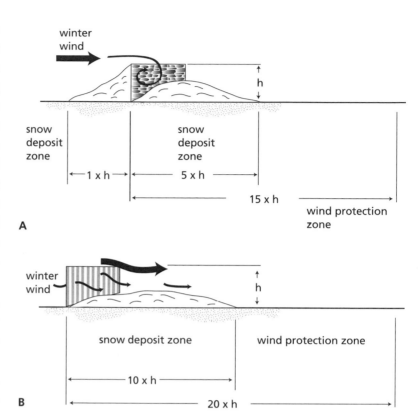

A solid windbreak (A) breaks the force of the wind less effectively than a porous structure (B), which creates a greater wind protection zone. This is especially critical in snow country, where solid structures gather snow on the upwind side and deep drifts on the downwind side.

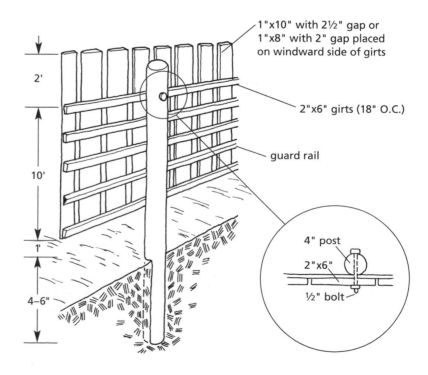

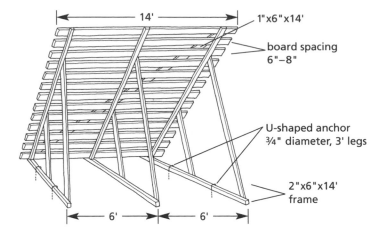

14'

1"x6"x14'

board spacing
6"–8"

U-shaped anchor
¾" diameter, 3' legs

2"x6"x14'
frame

6'   6'

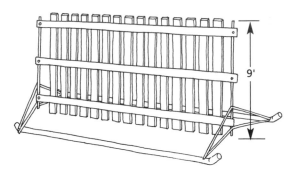

9'

There are dozens of designs for portable windbreak fences. Portable windbreaks should provide 20 percent openings, just like permanent windbreak fences.

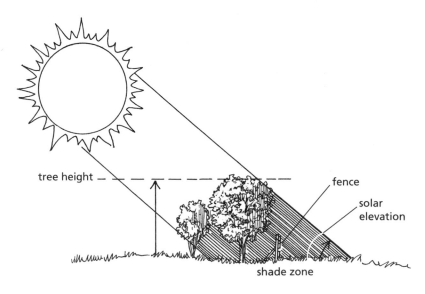

tree height

fence

solar elevation

shade zone

Trees can provide shade in pastures, but protect trees with a fence; animals having direct access to trees may damage their root systems. The area that receives shade is determined by the height of the trees and the elevation of the sun.

SHADE

In extreme summer heat, shade and plenty of fresh water work wonders in reducing body temperatures and keeping critters comfortable. Don't allow animals to camp in the shelterbelt because they can kill the trees. By carefully placing a fence along the edge of the shelterbelt, you allow animals to take advantage of the shade zone during the heat of day while protecting the trees and encouraging grazing during the rest of the day. Portable electric fencing works well for this application because you can keep animals completely out of the shaded areas in cool weather, when they shouldn't be spending time there, but allow them some access during times when heat stress is a legitimate concern.

The strategic placement of a few deciduous trees in a pasture or near buildings (remember, *near* doesn't mean adjacent to the buildings — allow 50 feet between trees and a building) provides cool refuges and, in the latter case, reduces in-building temperatures. Clumps of two or three trees work better than a single tree; fence off the clumps to protect them, just as you would a shelterbelt. Animals can follow the shade throughout the day, as needed, and growing trees are protected. Avoid planting trees due south of a building you are trying to protect: because of the changing angle of the sun throughout the year, trees directly south of a structure cannot block peak-summer sun when it's at its zenith, but they will block the sun's warming rays when it's low in the sky in midwinter. As such, it's best to plant clumps of trees to the southeast and southwest of a building.

Permanent or portable shade structures are also a great way to provide shade, particularly if you are raising hogs on pasture, as they are among the most heat-sensitive species, or if you are using rotational grazing for other species. One problem with permanent shade structures, however, is the resulting concentration of manure in the area of the structure;

nevertheless, they might be the best choice for some applications. Permanent structures should be at least 6 feet high on the back wall (though 8 feet is even better) and have their long axis oriented east to west. Use corrugated metal on the roof and paint the outside white and the inside black. For even better cooling, layer about 6 inches of straw on the roof and hold it down with chicken wire. Permanent structures can be designed to provide wind protection in winter and shade in summer.

Portable structures offer flexibility and greater freedom from problems associated with manure accumulation. And because they can be moved as part of a rotational system, they also encourage better utilization of forage. These structures can be constructed using wood and metal, but many people are now using shade cloth (sold by greenhouse manufacturers; see resources) on metal-tube or PVC frames; see page 84 for illustration.

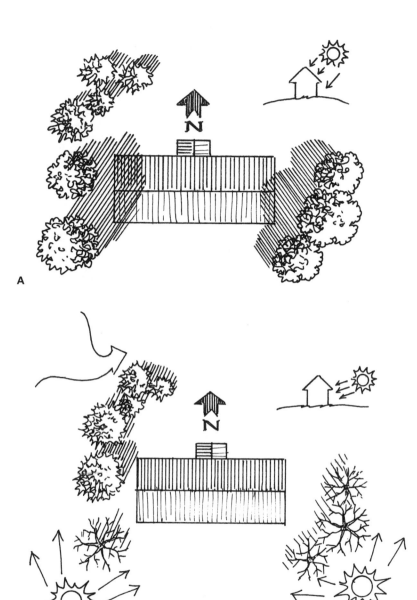

Trees planted near buildings can serve several functions. Deciduous trees planted at the southeast and southwest corners provide shade during summer but allow winter sun to strike the building (A). Evergreen trees to the northwest provide some shade in summer but, more important, provide a winter windbreak (B).

RECOMMENDED SPACE FOR SHADE/SHELTER

Stage of Production	Shade/Shelter
Growing finishing pigs	4 sq. ft./pig to 100 lbs., or 6 sq. ft./pig over 100 lbs.
Sows	15–20 sq. ft./sow
Sows and litters	20–30 sq. ft./sow and litter
Boars	40 –60 sq. ft./boar
Calves	20–25 sq. ft./calf to 400 lbs.
Feeders	30–35 sq. ft./feeder to 800 lbs.
Beef cows	35–40 sq. ft./cow
Dairy cows	45–50 sq. ft./cow
Bulls	45–50 sq. ft./bull
Chickens	1–2 sq. ft./bird
Ducks/geese	2–3 sq. ft./bird
Turkeys	2–4 sq. ft./bird

Permanent Shade Structure

This design can be built as a permanent shade structure or on skids (see side elevation and page 82). The metal roofing on this design *must* be screwed securely to the purlins, or it will pull away. (See page 244 for plan credit.)

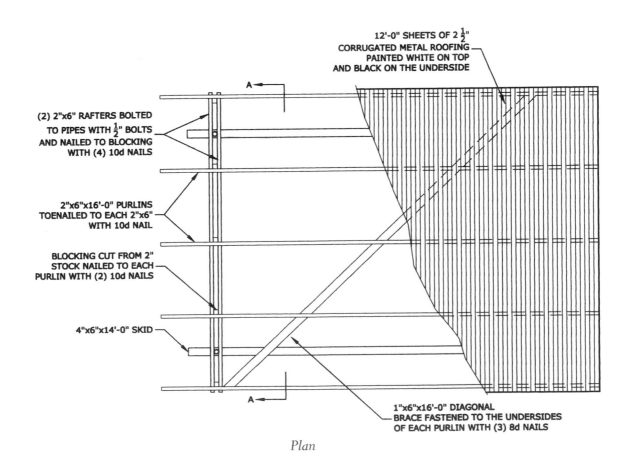

12'-0" SHEETS OF 2½"
CORRUGATED METAL ROOFING
PAINTED WHITE ON TOP
AND BLACK ON THE UNDERSIDE

(2) 2"x6" RAFTERS BOLTED
TO PIPES WITH ½" BOLTS
AND NAILED TO BLOCKING
WITH (4) 10d NAILS

2"x6"x16'-0" PURLINS
TOENAILED TO EACH 2"x6"
WITH 10d NAIL

BLOCKING CUT FROM 2"
STOCK NAILED TO EACH
PURLIN WITH (2) 10d NAILS

4"x6"x14'-0" SKID

1"x6"x16'-0" DIAGONAL
BRACE FASTENED TO THE UNDERSIDES
OF EACH PURLIN WITH (3) 8d NAILS

Plan

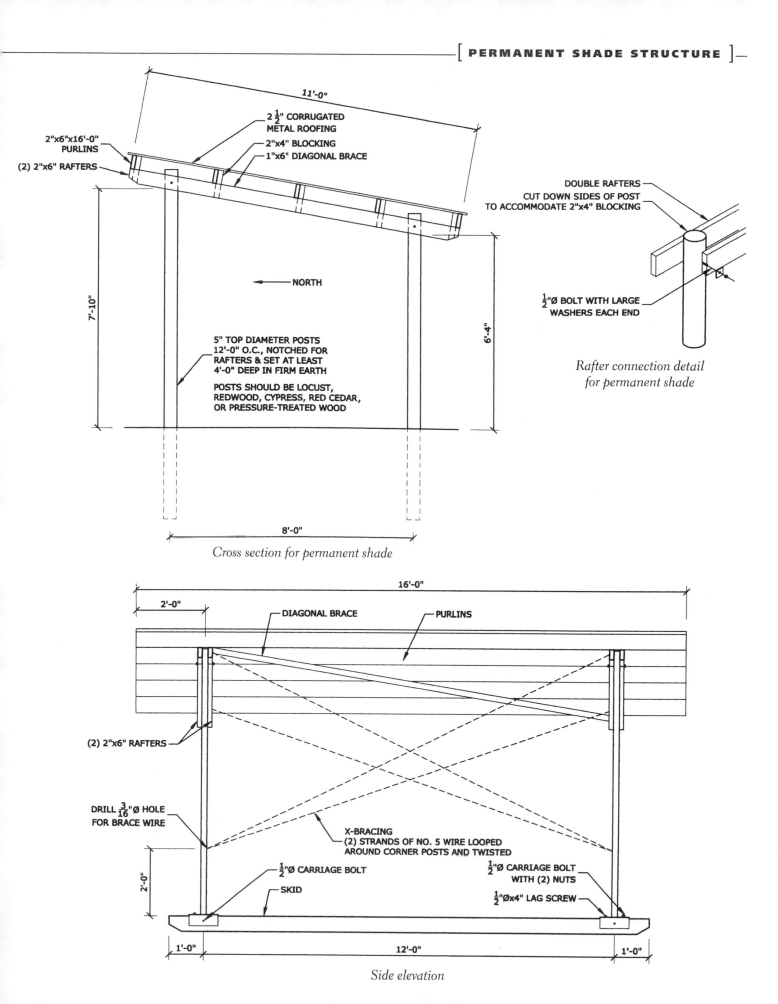

11'-0"

2 ½" CORRUGATED
METAL ROOFING

2"x6"x16'-0"
PURLINS

2"x4" BLOCKING

1"x6" DIAGONAL BRACE

(2) 2"x6" RAFTERS

DOUBLE RAFTERS

CUT DOWN SIDES OF POST
TO ACCOMMODATE 2"x4" BLOCKING

7'-10"

6'-4"

½"Ø BOLT WITH LARGE
WASHERS EACH END

NORTH

5" TOP DIAMETER POSTS
12'-0" O.C., NOTCHED FOR
RAFTERS & SET AT LEAST
4'-0" DEEP IN FIRM EARTH

POSTS SHOULD BE LOCUST,
REDWOOD, CYPRESS, RED CEDAR,
OR PRESSURE-TREATED WOOD

*Rafter connection detail
for permanent shade*

8'-0"

Cross section for permanent shade

16'-0"

2'-0"

DIAGONAL BRACE

PURLINS

(2) 2"x6" RAFTERS

DRILL 3/16"Ø HOLE
FOR BRACE WIRE

X-BRACING
(2) STRANDS OF NO. 5 WIRE LOOPED
AROUND CORNER POSTS AND TWISTED

½"Ø CARRIAGE BOLT

½"Ø CARRIAGE BOLT
WITH (2) NUTS

2'-0"

SKID

½"Ø x 4" LAG SCREW

1'-0"

12'-0"

1'-0"

Side elevation

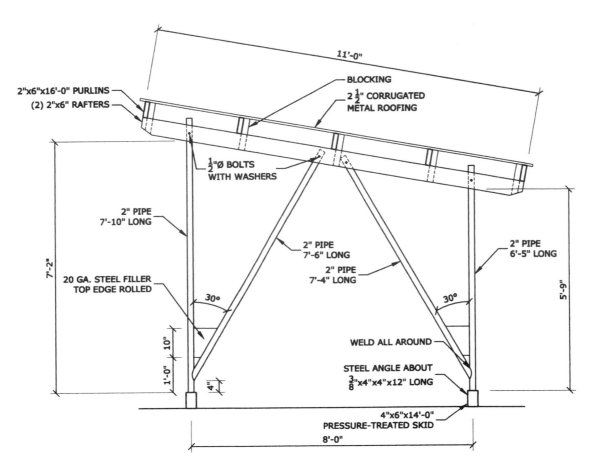

11'-0"

2"x6"x16'-0" PURLINS
(2) 2"x6" RAFTERS

BLOCKING
2½" CORRUGATED
METAL ROOFING

½"Ø BOLTS
WITH WASHERS

2" PIPE
7'-10" LONG

2" PIPE
7'-6" LONG

2" PIPE
7'-4" LONG

2" PIPE
6'-5" LONG

7'-2"

20 GA. STEEL FILLER
TOP EDGE ROLLED

30°

30°

5'-9"

WELD ALL AROUND

STEEL ANGLE ABOUT
⅜"x4"x4"x12" LONG

1'-0" 10"

4"

4"x6"x14'-0"
PRESSURE-TREATED SKID

8'-0"

Cross section at A-A for portable shade

Ø = diameter
GA. = gauge
O.C. = on center

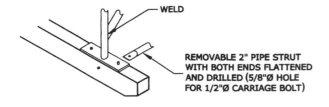

WELD

REMOVABLE 2" PIPE STRUT
WITH BOTH ENDS FLATTENED
AND DRILLED (5/8"Ø HOLE
FOR 1/2"Ø CARRIAGE BOLT)

*Skid connector detail
(struts are removed after moving)*

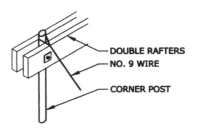

DOUBLE RAFTERS
NO. 9 WIRE
CORNER POST

Cross-bracing detail

Combination Windbreak & Shade

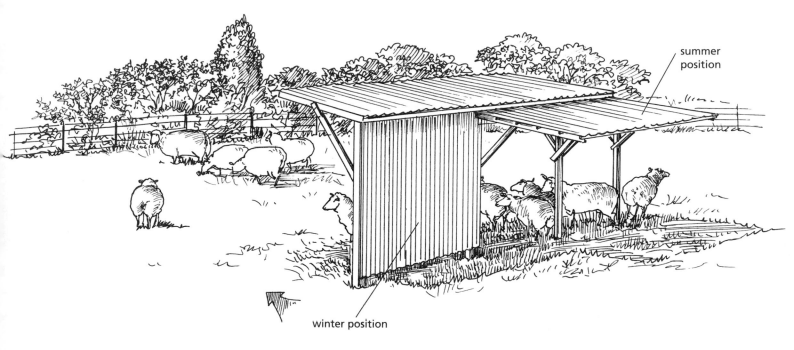

summer position

winter position

This permanent structure is wonderful because it functions as both
a windbreak and a shade structure. In winter, the side is lowered to shield
animals; in summer it is raised, extending the shaded area and improving
ventilation. (See page 245 for plan credit.)

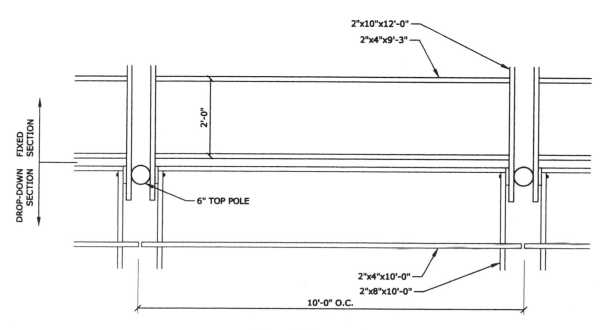

2"x10"x12'-0"

2"x4"x9'-3"

2'-0"

6" TOP POLE

FIXED SECTION

DROP-DOWN SECTION

2"x4"x10'-0"

2"x8"x10'-0"

10'-0" O.C.

Plan of pole connections

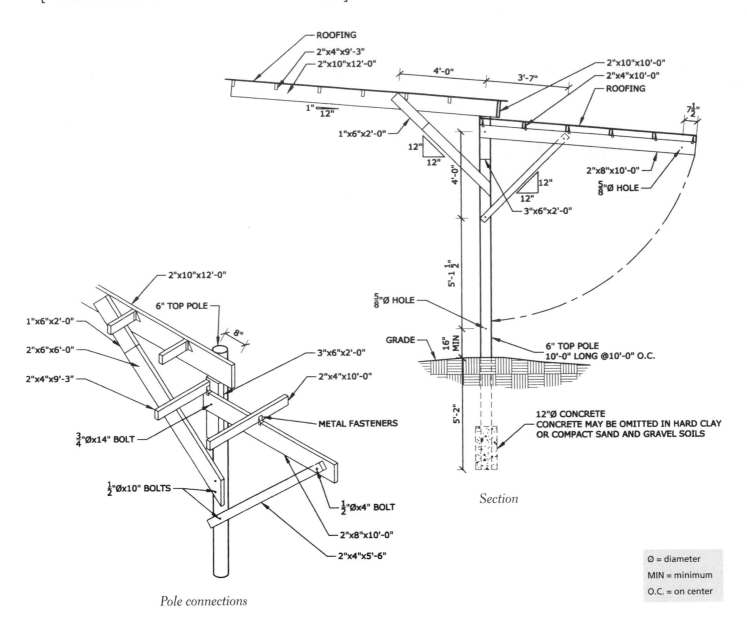

ROOFING
2"x4"x9'-3"
2"x10"x12'-0"
2"x10"x10'-0"
2"x4"x10'-0"
ROOFING
4'-0"
3'-7"
$7\frac{1}{2}$"
$\frac{1}{12}$"
1"x6"x2'-0"
12"
12"
4'-0"
12"
12"
2"x8"x10'-0"
$\frac{5}{8}$"Ø HOLE
3"x6"x2'-0"
5'-1$\frac{1}{2}$"
$\frac{5}{8}$"Ø HOLE
GRADE
16" MIN
6" TOP POLE
10'-0" LONG @10'-0" O.C.
5'-2"
12"Ø CONCRETE
CONCRETE MAY BE OMITTED IN HARD CLAY
OR COMPACT SAND AND GRAVEL SOILS

Section

2"x10"x12'-0"
6" TOP POLE
8"
1"x6"x2'-0"
2"x6"x6'-0"
2"x4"x9'-3"
3"x6"x2'-0"
2"x4"x10'-0"
$\frac{3}{4}$"Øx14" BOLT
METAL FASTENERS
$\frac{1}{2}$"Øx10" BOLTS
$\frac{1}{2}$"Øx4" BOLT
2"x8"x10'-0"
2"x4"x5'-6"

Pole connections

Ø = diameter
MIN = minimum
O.C. = on center

Shade Cloth Structure

Shade cloth structures can be
created with shade cloth
(available at greenhouse
suppliers) or tarps. Tying down
the cloth securely is key. This
example was created by using stock
panels and wiring curved metal
tubing to the panels to support the roof.

Portable Shade Structure

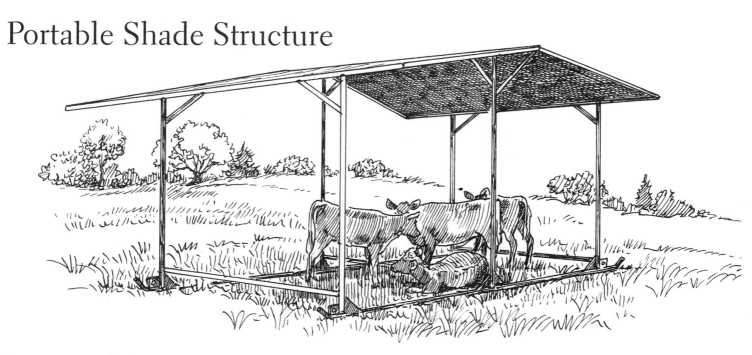

The measurements for this portable structure are flexible,
but the services of a welder are required during assembly. Because
it can be built as a tall structure (up to 13 feet at the peak), it is suitable
for draft horses. (See page 245 for plan credit.)

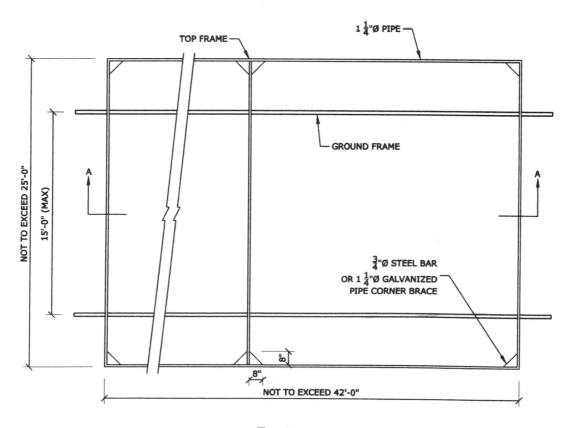

Top view

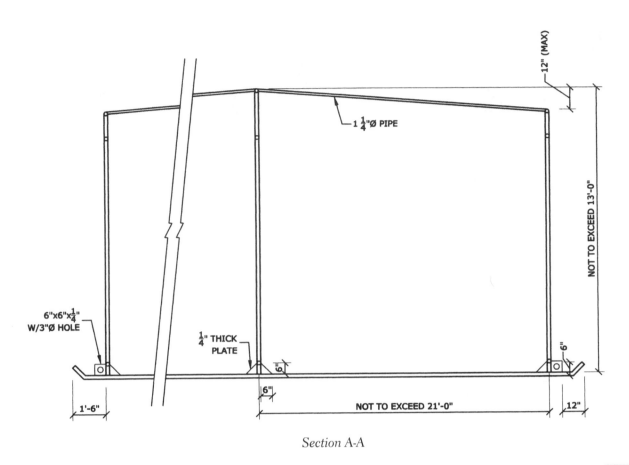

12" (MAX)

1 $\frac{1}{4}$"Ø PIPE

NOT TO EXCEED 13'-0"

6"x6"x$\frac{1}{4}$"
W/3"Ø HOLE

$\frac{1}{4}$" THICK
PLATE

6"

6"

6"

1'-6"

6"

NOT TO EXCEED 21'-0"

12"

Section A-A

Ø = diameter

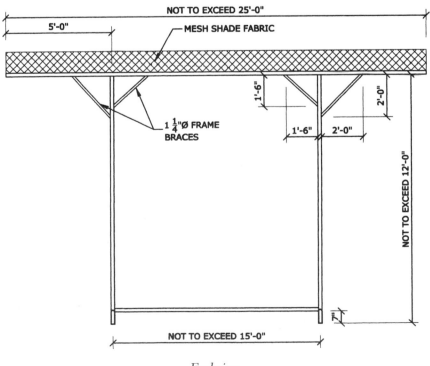

NOT TO EXCEED 25'-0"

5'-0"

MESH SHADE FABRIC

1 $\frac{1}{4}$"Ø FRAME
BRACES

1'-6"

2'-0"

1'-6"

2'-0"

NOT TO EXCEED 12'-0"

7"

NOT TO EXCEED 15'-0"

End view

Portable Shade for Hogs

This welded structure is suitable for pigs, sheep, goats, and small cattle. If you will use it in a windy area, be sure to tie it down. Drive in tent stakes at each corner. Throw a nylon rope over the roof and cut it so it almost spans the distance between the stakes at one end. Tie a bungee cord to each end of the rope, then secure the bungees to the stakes. Do the same at the other end. The bungee cords give a little in high winds and will keep the rope from wearing out prematurely. (See page 245 for plan credit.)

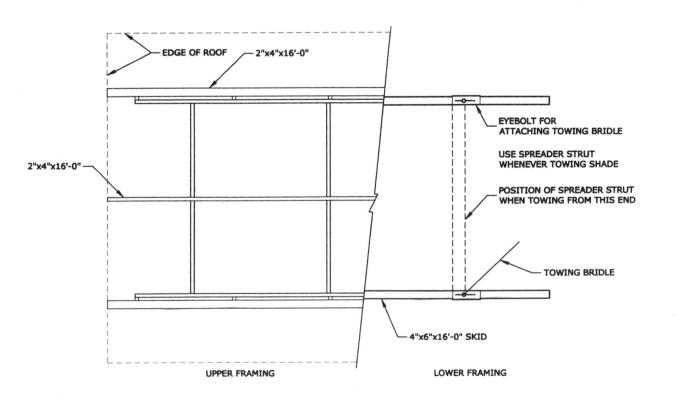

Plan

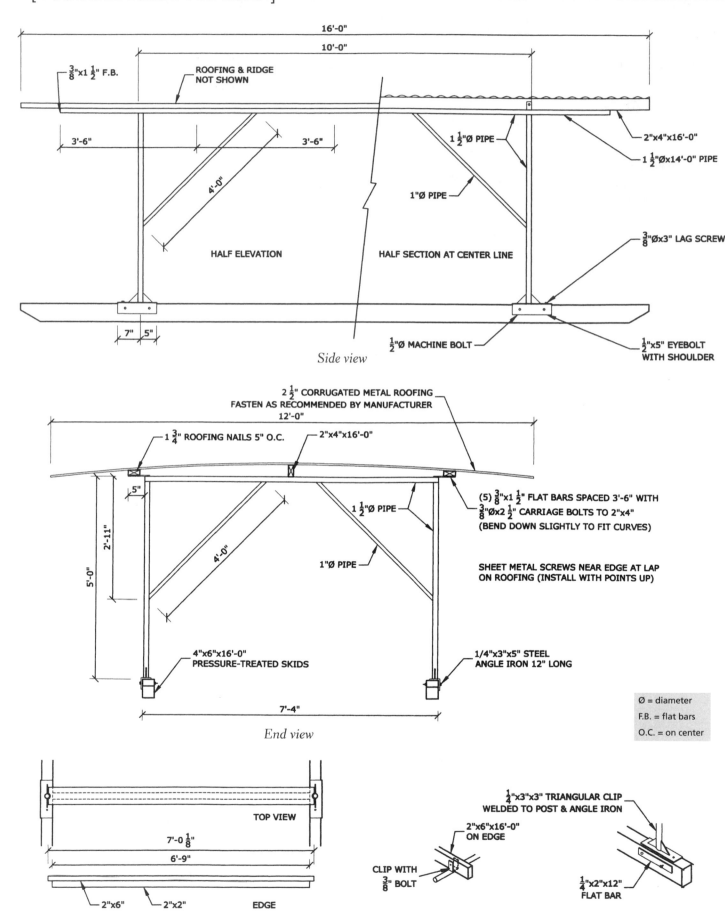

16'-0"

10'-0"

$\frac{3}{8}$"x1 $\frac{1}{2}$" F.B.

ROOFING & RIDGE
NOT SHOWN

3'-6"

3'-6"

4'-0"

HALF ELEVATION

1 $\frac{1}{2}$"Ø PIPE

1"Ø PIPE

HALF SECTION AT CENTER LINE

2"x4"x16'-0"

1 $\frac{1}{2}$"Øx14'-0" PIPE

$\frac{3}{8}$"Øx3" LAG SCREW

7" 5"

$\frac{1}{2}$"Ø MACHINE BOLT

$\frac{1}{2}$"x5" EYEBOLT
WITH SHOULDER

Side view

2 $\frac{1}{2}$" CORRUGATED METAL ROOFING
FASTEN AS RECOMMENDED BY MANUFACTURER

12'-0"

1 $\frac{3}{4}$" ROOFING NAILS 5" O.C.

2"x4"x16'-0"

5"

2'-11"

5'-0"

1 $\frac{1}{2}$"Ø PIPE

4'-0"

1"Ø PIPE

(5) $\frac{3}{8}$"x1 $\frac{1}{2}$" FLAT BARS SPACED 3'-6" WITH
$\frac{3}{8}$"Øx2 $\frac{1}{2}$" CARRIAGE BOLTS TO 2"x4"
(BEND DOWN SLIGHTLY TO FIT CURVES)

SHEET METAL SCREWS NEAR EDGE AT LAP
ON ROOFING (INSTALL WITH POINTS UP)

4"x6"x16'-0"
PRESSURE-TREATED SKIDS

1/4"x3"x5" STEEL
ANGLE IRON 12" LONG

7'-4"

End view

Ø = diameter
F.B. = flat bars
O.C. = on center

TOP VIEW

7'-0 $\frac{1}{8}$"

6'-9"

2"x6"   2"x2"   EDGE

Spreader strut

$\frac{1}{4}$"x3"x3" TRIANGULAR CLIP
WELDED TO POST & ANGLE IRON

2"x6"x16'-0"
ON EDGE

CLIP WITH
$\frac{3}{8}$" BOLT

$\frac{1}{4}$"x2"x12"
FLAT BAR

Skid connection

6. BARNS & STABLES

Ready to tackle a barn or stable project? The barns and sheds described in this chapter are good general-purpose structures suitable for one or more species of livestock. They are adaptable, in that floor plans can be adjusted to meet personal needs, and they are based on proven designs, many of which have been fine-tuned over decades by various land-grant universities and by the U.S. Department of Agriculture. These are structures that will create useful space in your backyard, on your farm, or on your ranch. The plans here are generally considered conceptual; they are intended to provide ideas, though there are hundreds of other design options around. They're not working drawings — that is, the drawings have not been finalized according to your local code and structural requirements

— and so probably won't pass muster with the local building department. An architect or engineer can develop working drawings based on these plans that meet specific code requirements, or you may be able to develop the working plans yourself by working with a county building official or a knowledgeable builder.

If you intend to do the work yourself but don't have much construction experience, I urge you to try some smaller projects first and to study the construction information in part III before tackling a project in this chapter. Even if you plan to hire a contractor to build your dream barn, reading the how-to chapters in part III will give you a solid understanding of the process and help you become an educated consumer.

TRADITIONAL GAMBREL-ROOFED BARNS

If you are thinking about a traditional style of barn, this is it. Gambrel-roofed barns epitomize our agricultural past but can easily satisfy the needs of contemporary users. We had a gambrel barn on our farm in Minnesota, and it served us well.

These barns are complicated to construct because of the roof system, but trusses can be purchased to make the project easier. Because of their size, full-size gambrel barns usually require a concrete-wall foundation with interior posts. They may include concrete slabs, particularly for tack and feed rooms. They also call for wind bracing in the walls, which is achieved by placing boards on an angle along the inside of walls.

In areas with moderate to low snow loads and moderate wind loads, this design calls for 2x6 lumber for trusses and studs, 6x6 posts for support of the second story, and 2x10 girders and joists that support the second-story haymow. If you plan to adapt the design for high snow or wind zones, all these will need to be sized up to meet local load demands, and the mow may require additional collars run between the trusses at approximately one-third the height from the top of the peak. The land-grant college in your state may offer a set of working plans for this style barn that will meet local snow loads. Ask your Extension agent about the availability of locally adapted plans.

Gambrel Barn

This is the traditional dairy-barn style common throughout
the Midwest dairy region and seen sporadically in other areas
of the United States. Usually painted red with white trim and topped
by a cupola, it is quintessential country. (See page 245 for plan credits.)

FEED ROOM

LARGE BOX STALL

6"x6" POST

REMOVABLE PANEL

LARGE BOX STALL

HAY DROP

DRIVE THRU ALLEY

HAY DROP

TACK & STORAGE

PROVIDE STAIRS TO LOFT

STALL

STALL

STALL

Floor plan

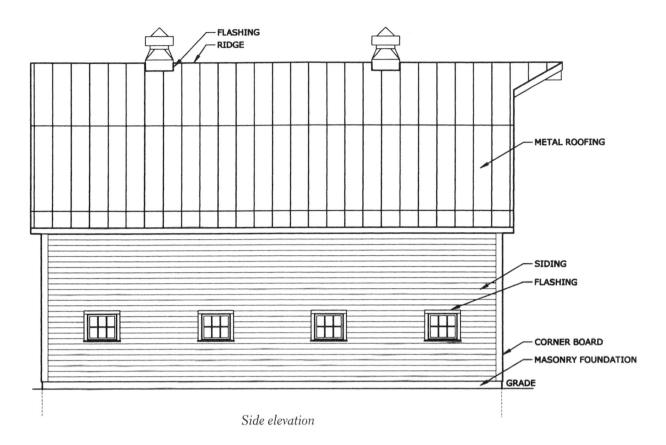

FLASHING
RIDGE

METAL ROOFING

SIDING
FLASHING

CORNER BOARD
MASONRY FOUNDATION
GRADE

Side elevation

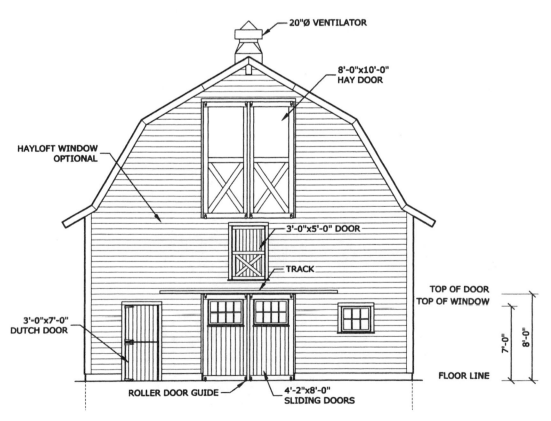

20"Ø VENTILATOR

8'-0"x10'-0"
HAY DOOR

HAYLOFT WINDOW
OPTIONAL

3'-0"x5'-0" DOOR

TRACK

TOP OF DOOR
TOP OF WINDOW

7'-0"   8'-0"

3'-0"x7'-0"
DUTCH DOOR

FLOOR LINE

ROLLER DOOR GUIDE

4'-2"x8'-0"
SLIDING DOORS

End elevation

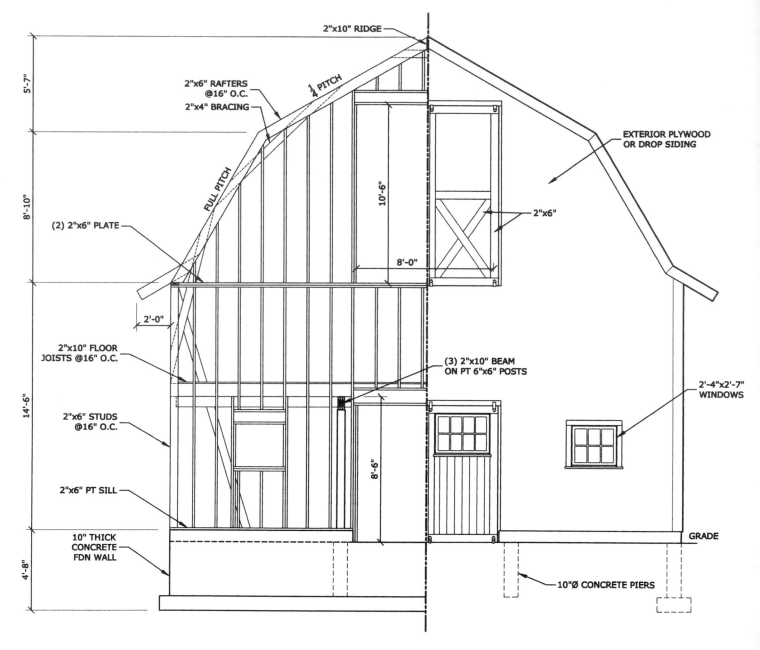

2"x10" RIDGE

2"x6" RAFTERS @16" O.C.

2"x4" BRACING

¼ PITCH

FULL PITCH

EXTERIOR PLYWOOD OR DROP SIDING

(2) 2"x6" PLATE

10'-6"

2"x6"

8'-0"

5'-7"

8'-10"

2'-0"

2"x10" FLOOR JOISTS @16" O.C.

(3) 2"x10" BEAM ON PT 6"x6" POSTS

2'-4"x2'-7" WINDOWS

2"x6" STUDS @16" O.C.

8'-6"

14'-6"

2"x6" PT SILL

GRADE

10" THICK CONCRETE FDN WALL

10"Ø CONCRETE PIERS

4'-8"

Partial framing, end elevation

Gambrel Roofs

Gambrel roofs have long been a standard design for barns. A true gambrel roof yields a large area of usable space in the second story, which traditionally served as hay storage. Today, this area can provide hay storage or guest/employee housing, an office, a studio, or a meeting area.

For small sheds, gambrel-style roofs are generally built with a gambrel truss. It doesn't yield much extra room, but it can be used when the style is desirable for aesthetic reasons.

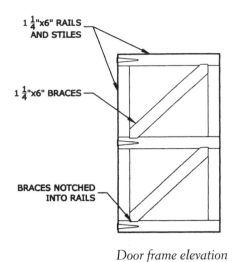

1 ¼"x6" RAILS AND STILES

1 ¼"x6" BRACES

BRACES NOTCHED INTO RAILS

Door frame elevation

EXTERIOR FINISH

EXTERIOR TRIM WITH FLASHING

WINDOW STOPS

WINDOW IN CLOSED POSITION

WINDOWSILL

(2) 2"x6"

2"x6" STUDS

(2) 2"X8" BOX HEADER

WINDOW IN FULL OPEN POSITION

1"x2" WOOD TIE BETWEEN METAL WINDOW GUARDS

PEG INSERT HOLES FOR WINDOW ADJUSTMENT

METAL WINDOW GUARDS SECURED TO WINDOW JAMBS

WINDOW STOOL WITH TRIM

Section

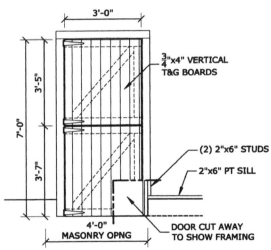

3'-0"

3'-5"

7'-0"

3'-7"

4'-0" MASONRY OPNG

¾"x4" VERTICAL T&G BOARDS

(2) 2"x6" STUDS

2"x6" PT SILL

DOOR CUT AWAY TO SHOW FRAMING

Exterior elevation, Dutch door

Ø = diameter
FDN = foundation
O.C. = on center
OPNG = opening
PT = pressure-treated
T&G = tongue and groove

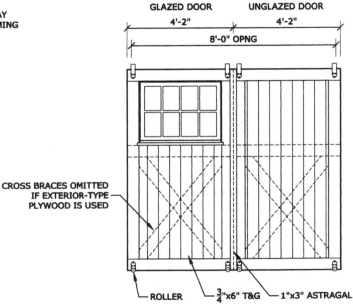

GLAZED DOOR
4'-2"

UNGLAZED DOOR
4'-2"

8'-0" OPNG

CROSS BRACES OMITTED IF EXTERIOR-TYPE PLYWOOD IS USED

ROLLER

¾"x6" T&G

1"x3" ASTRAGAL

Exterior elevation, double sliding door

Gambrel-Style Small Barn

Based on designs similar to the
gambrel barn, this small version is ideal for occasional
use for livestock and for feed and tack storage. Thanks to its
reduced size, this barn can be constructed either on a pier or post foundation
or on a concrete slab on footer foundation system. Concrete floors are hard on an animal's
legs, so if you are using a slab for the entire structure, it should have a wooden layer, rubber
tiles or mats, or a deep layer of bedding within the box stalls. Six-by-six posts in the center
support a girder system holding the mow floor. (See page 245 for plan credits.)

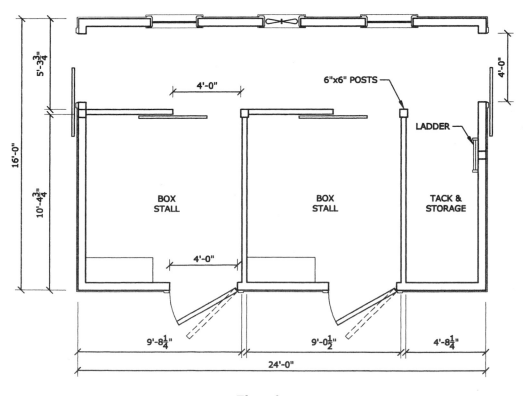

Floor plan

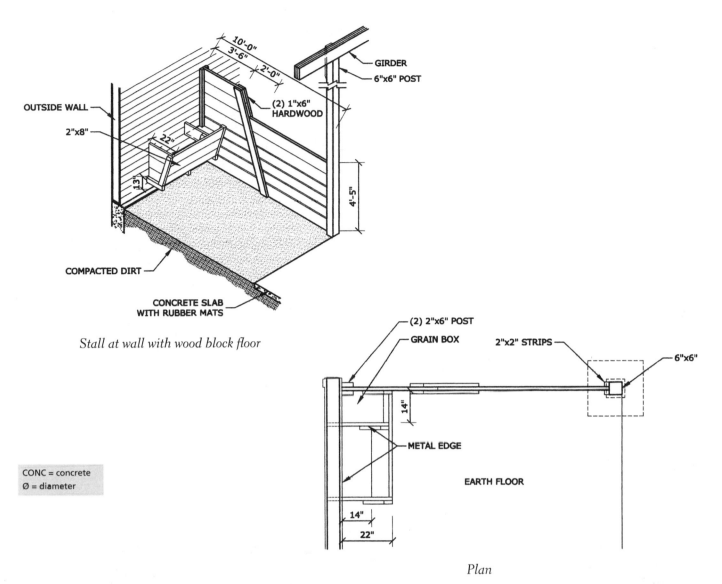

OUTSIDE WALL

2"x8"

22"

13"

COMPACTED DIRT

CONCRETE SLAB
WITH RUBBER MATS

10'-0"
3'-6" 2'-0"

(2) 1"x6"
HARDWOOD

GIRDER
6"x6" POST

4'-5"

Stall at wall with wood block floor

CONC = concrete
Ø = diameter

(2) 2"x6" POST
GRAIN BOX
2"x2" STRIPS
6"x6"

METAL EDGE

EARTH FLOOR

14"

14"
22"

Plan

GIRDER
BLOCK
6"x6" POST

(2) 2"x6" PLATE

3'-0" 2'-0"

2"x2" STRIPS

22"

2"x10"
2"x8"

2"x4"

4'-5"

6'-0"

3'-5"

13"

14"

2"x10"

EARTH

FOUNDATION WALL

RUBBER MAT ON
CONCRETE SLAB

8"Ø CONC PIER
ON FOOTING

Section

BASIC GABLE-ROOFED BARNS

Gable-roofed structures are simpler to build than gambrel-style barns and can be very attractive. These structures can be built on pier and beam, post, or concrete slab and footer foundations, depending on final interior design. The upper story can be fully enclosed to provide hay storage, it can be designed with a partial loft for storage but with an airier feeling, or it can be open with skylights for the airiest feel of all. In the loft and open versions, ventilation is easier to maintain, whereas an enclosed second story will require some extra planning for ventilation. Gable designs can be stick built with studs or based on traditional post-barn construction techniques.

Small Gable Barn

This is a great little barn that evokes the feel of a small New England village, but with its basic lines and great proportions, it would fit well in any part of the country. Thanks to the relatively steep pitch of its roof, it sheds snow well and is therefore an attractive design for high snowfall areas. (See page 245 for plan credit.)

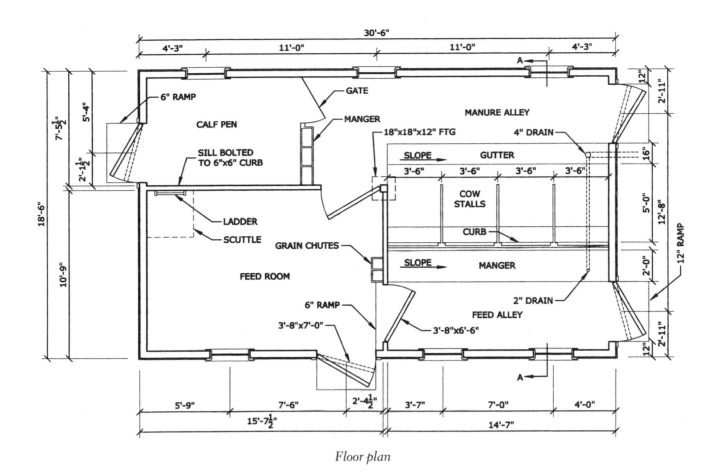

Floor plan

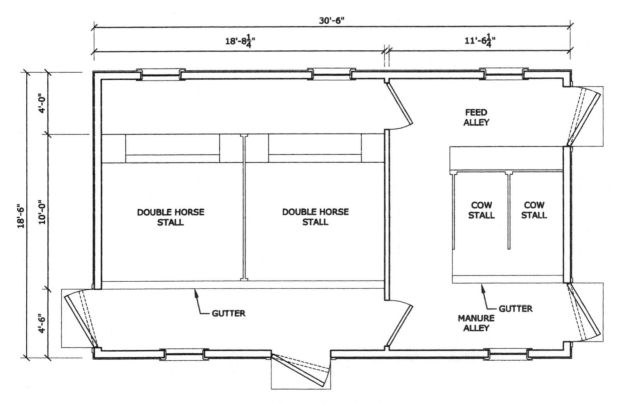

Alternate floor plan 1

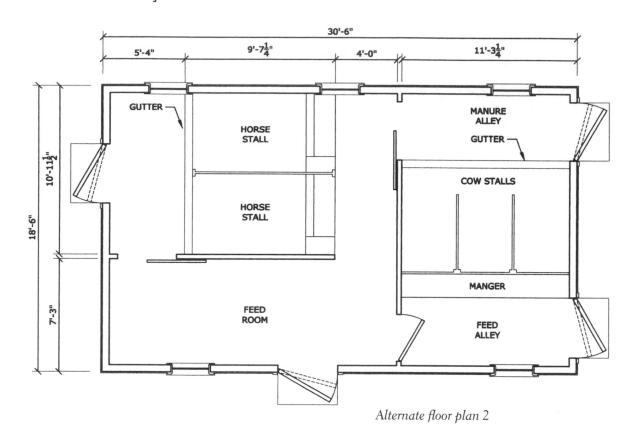

GUTTER

HORSE STALL

HORSE STALL

FEED ROOM

MANURE ALLEY

GUTTER

COW STALLS

MANGER

FEED ALLEY

30'-6"

5'-4"   9'-7¼"   4'-0"   11'-3¼"

10'-11½"

18'-6"

7'-3"

Alternate floor plan 2

FTG = footing
O.C. = on center

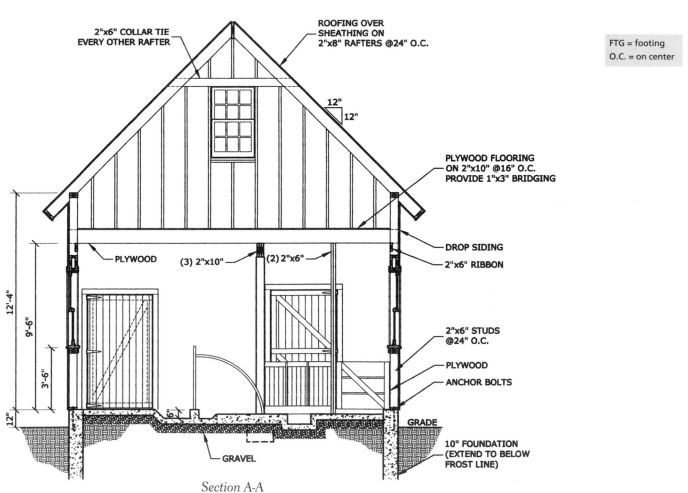

2"x6" COLLAR TIE EVERY OTHER RAFTER

ROOFING OVER SHEATHING ON 2"x8" RAFTERS @24" O.C.

12"
12"

PLYWOOD FLOORING ON 2"x10" @16" O.C. PROVIDE 1"x3" BRIDGING

DROP SIDING

2"x6" RIBBON

PLYWOOD

(3) 2"x10"   (2) 2"x6"

2"x6" STUDS @24" O.C.

PLYWOOD

ANCHOR BOLTS

12'-4"

9'-6"

3'-6"

12"

6"

GRADE

GRAVEL

10" FOUNDATION (EXTEND TO BELOW FROST LINE)

Section A-A

Medium Gable Barn

This barn is similar to the small gable barn but larger. The plan was originally developed in North Carolina, so for northern areas, increase to 2x6 or 2x8 walls and rafters and 6x6 columns. (See page 245 for plan credit.)

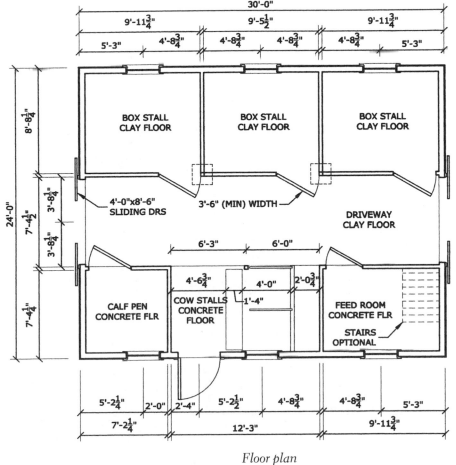

BOX STALL CLAY FLOOR

BOX STALL CLAY FLOOR

BOX STALL CLAY FLOOR

4'-0"x8'-6" SLIDING DRS

3'-6" (MIN) WIDTH

DRIVEWAY CLAY FLOOR

CALF PEN CONCRETE FLR

COW STALLS CONCRETE FLOOR

FEED ROOM CONCRETE FLR

STAIRS OPTIONAL

Floor plan

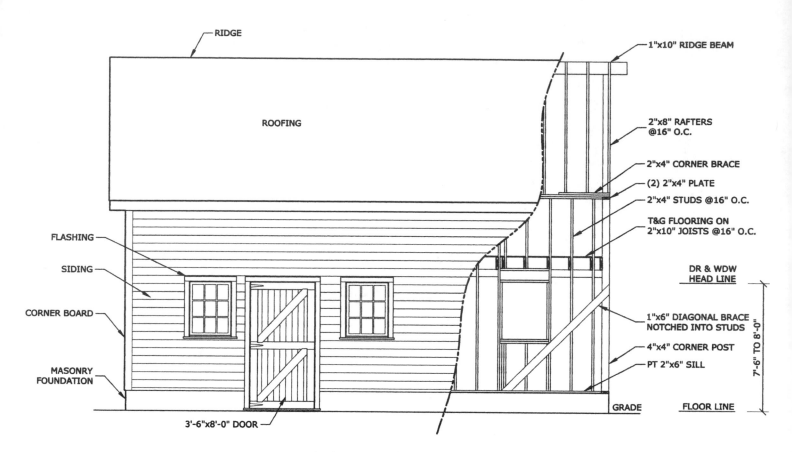

RIDGE

1"x10" RIDGE BEAM

ROOFING

2"x8" RAFTERS @16" O.C.

2"x4" CORNER BRACE

(2) 2"x4" PLATE

2"x4" STUDS @16" O.C.

T&G FLOORING ON 2"x10" JOISTS @16" O.C.

FLASHING

SIDING

CORNER BOARD

MASONRY FOUNDATION

DR & WDW HEAD LINE

1"x6" DIAGONAL BRACE NOTCHED INTO STUDS

4"x4" CORNER POST

PT 2"x6" SILL

GRADE

FLOOR LINE

7'-6" TO 8'-0"

3'-6"x8'-0" DOOR

Side elevation and partial framing

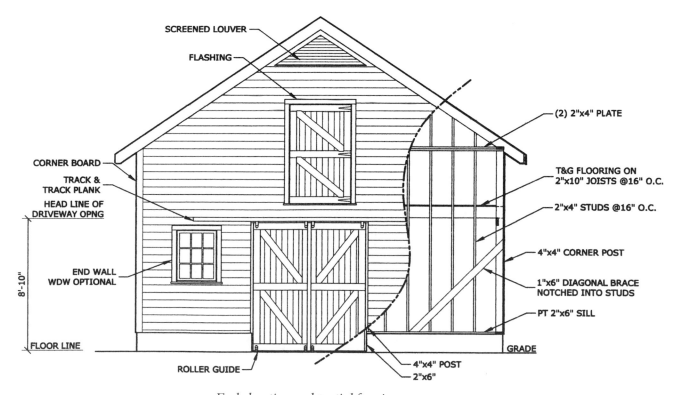

SCREENED LOUVER

FLASHING

CORNER BOARD

TRACK & TRACK PLANK

HEAD LINE OF DRIVEWAY OPNG

END WALL WDW OPTIONAL

8'-10"

FLOOR LINE

ROLLER GUIDE

4"x4" POST

2"x6"

(2) 2"x4" PLATE

T&G FLOORING ON 2"x10" JOISTS @16" O.C.

2"x4" STUDS @16" O.C.

4"x4" CORNER POST

1"x6" DIAGONAL BRACE NOTCHED INTO STUDS

PT 2"x6" SILL

GRADE

End elevation and partial framing

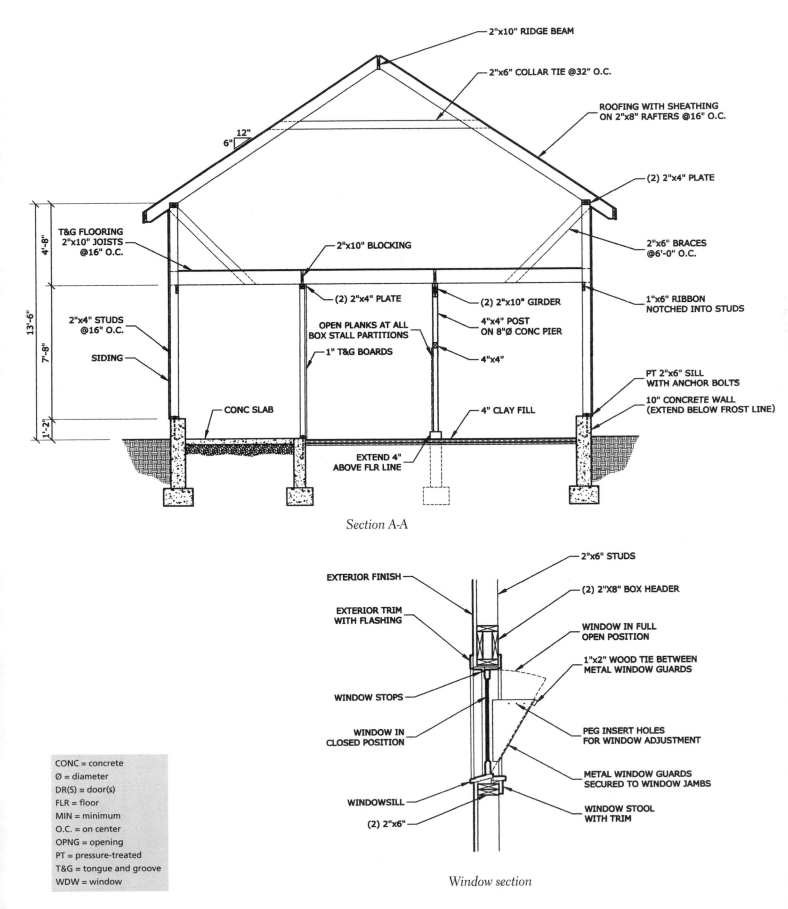

2"x10" RIDGE BEAM

2"x6" COLLAR TIE @32" O.C.

ROOFING WITH SHEATHING
ON 2"x8" RAFTERS @16" O.C.

6" 12"

(2) 2"x4" PLATE

2"x6" BRACES
@6'-0" O.C.

T&G FLOORING
2"x10" JOISTS
@16" O.C.

2"x10" BLOCKING

(2) 2"x4" PLATE

(2) 2"x10" GIRDER

1"x6" RIBBON
NOTCHED INTO STUDS

4'-8"

2"x4" STUDS
@16" O.C.

4"x4" POST
ON 8"Ø CONC PIER

OPEN PLANKS AT ALL
BOX STALL PARTITIONS

13'-6"

7'-8"

SIDING

1" T&G BOARDS

4"x4"

PT 2"x6" SILL
WITH ANCHOR BOLTS

10" CONCRETE WALL
(EXTEND BELOW FROST LINE)

1'-2"

CONC SLAB

4" CLAY FILL

EXTEND 4"
ABOVE FLR LINE

Section A-A

2"x6" STUDS

EXTERIOR FINISH

(2) 2"X8" BOX HEADER

EXTERIOR TRIM
WITH FLASHING

WINDOW IN FULL
OPEN POSITION

1"x2" WOOD TIE BETWEEN
METAL WINDOW GUARDS

WINDOW STOPS

WINDOW IN
CLOSED POSITION

PEG INSERT HOLES
FOR WINDOW ADJUSTMENT

METAL WINDOW GUARDS
SECURED TO WINDOW JAMBS

WINDOWSILL

(2) 2"x6"

WINDOW STOOL
WITH TRIM

CONC = concrete
Ø = diameter
DR(S) = door(s)
FLR = floor
MIN = minimum
O.C. = on center
OPNG = opening
PT = pressure-treated
T&G = tongue and groove
WDW = window

Window section

Monitor Barn

The monitor design is a combination of two gable-roof planes. Although the upper story can be fully enclosed, it is often fully open down the center or designed with lofts over either end of the barn. These latter configurations are open and airy, provide great light during daytime hours, and are easily ventilated. This is my dream barn! (See page 245 for plan credit.)

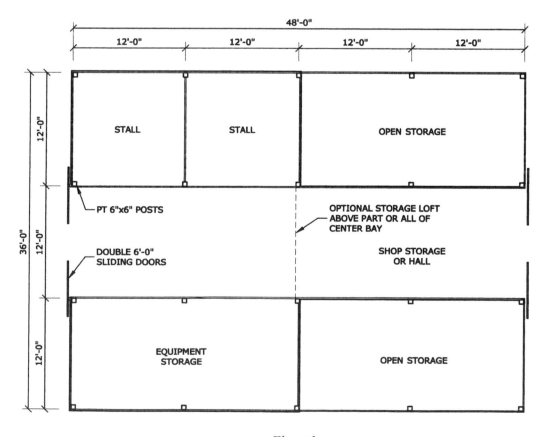

Floor plan

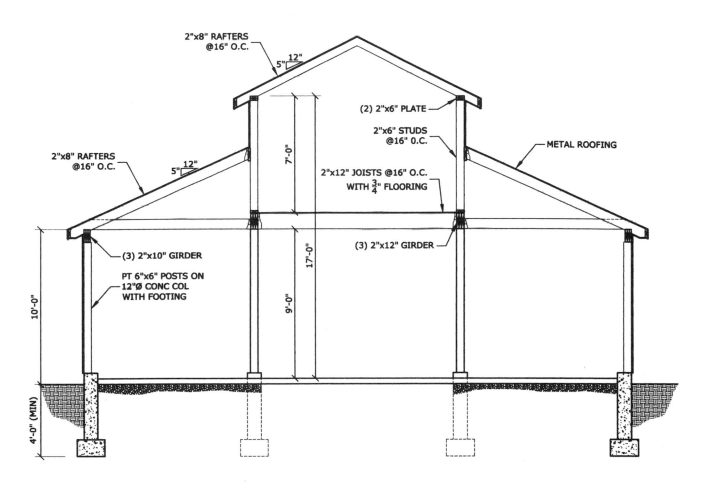

2"x8" RAFTERS
@16" O.C.

5" 12"

(2) 2"x6" PLATE

2"x6" STUDS
@16" O.C.

METAL ROOFING

2"x8" RAFTERS
@16" O.C.

5" 12"

2"x12" JOISTS @16" O.C.
WITH $\frac{3}{4}$" FLOORING

7'-0"

(3) 2"x10" GIRDER

(3) 2"x12" GIRDER

PT 6"x6" POSTS ON
12"Ø CONC COL
WITH FOOTING

17'-0"

9'-0"

10'-0"

4'-0" (MIN)

Cross section

Rear view

Front view

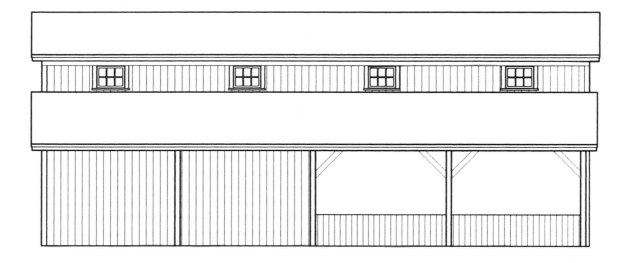

Right side view

CONC COL = concrete column
Ø = diameter
MIN = minimum
O.C. = on center
PT = pressure-treated

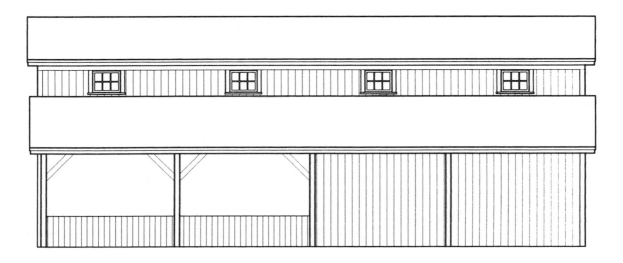

Left side view

SINGLE-STORY STABLES AND COOPS

For some reason, and I'm not sure what it is, single-story barns are usually called *stables*, unless they are intended for poultry, and then they're called *coops*. Although they're rarely called a barn, stables are usually appropriate for more than just horses (for example, one box stall of a stable can easily be converted to a chicken area), and chicken coops can often be adapted for other uses. A small backyard stable with a couple of box stalls is often all that's needed for one or two animals, but these single-story structures can also be quite large. In fact, in areas where race or show horses are raised in large numbers, like Kentucky and Florida, you can find these structures designed to house dozens and dozens of horses. Like two-story gable structures, they can be built using stud-wall construction or post-barn techniques; in hot climates, they are often built with concrete block walls.

Small Stable

This design is easy to construct and provides room for tack, tool, and feed storage as well as two good-sized box stalls. High windows on the back combined with Dutch doors on the front of each stall give great ventilation, and the overhanging porch is ideal for helping to keep it cool during hot weather; however, its relatively flat roof makes it a poor choice in high snowfall areas, where snow load exceeds 20 pounds per square foot. In higher snowfall areas, a steeper roof design with 2x6 rafters spaced on 16-inch centers can be substituted for the flat roof. Consult a local building official, architect, or engineer for advice on load and structural issues. (See page 246 for plan credit.)

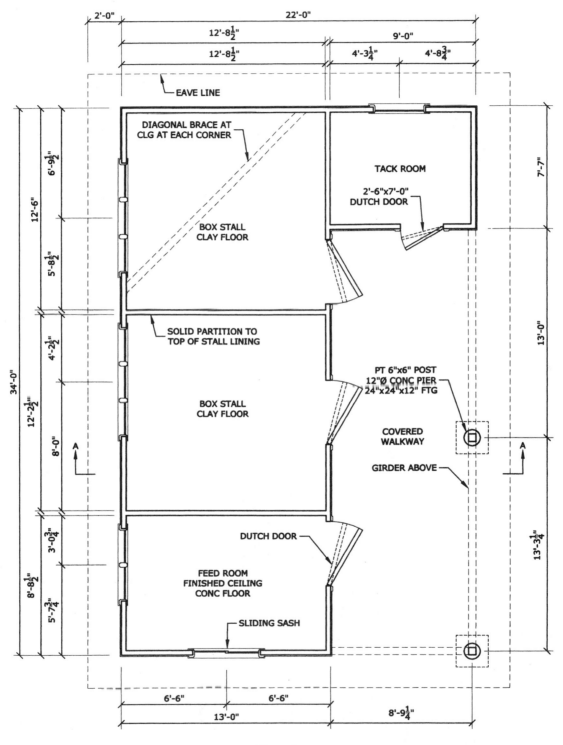

2'-0"
22'-0"
12'-8½"
9'-0"
12'-8½"
4'-3¼"
4'-8¾"

EAVE LINE

DIAGONAL BRACE AT
CLG AT EACH CORNER

TACK ROOM

2'-6"x7'-0"
DUTCH DOOR

BOX STALL
CLAY FLOOR

7'-7"

6'-9½"

12'-6"

5'-8½"

SOLID PARTITION TO
TOP OF STALL LINING

4'-2½"

PT 6"x6" POST
12"Ø CONC PIER
24"x24"x12" FTG

13'-0"

34'-0"

12'-2½"

BOX STALL
CLAY FLOOR

COVERED
WALKWAY

8'-0"

A

GIRDER ABOVE

A

DUTCH DOOR

3'-0¾"

13'-3¼"

8'-8½"

FEED ROOM
FINISHED CEILING
CONC FLOOR

5'-7¾"

SLIDING SASH

6'-6"
6'-6"
13'-0"
8'-9¼"

Floor plan

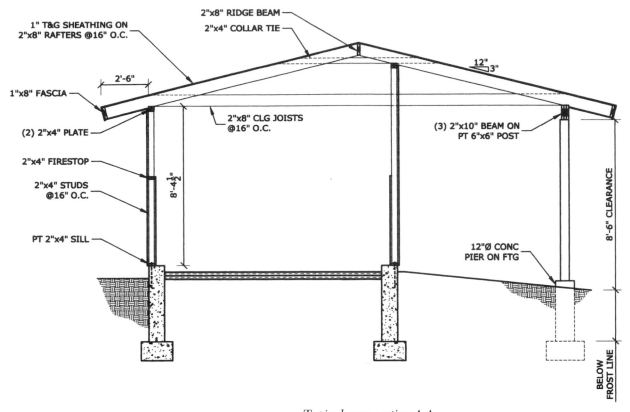

1" T&G SHEATHING ON
2"x8" RAFTERS @16" O.C.

2"x8" RIDGE BEAM

2"x4" COLLAR TIE

1"x8" FASCIA

2'-6"

12"
3"

(2) 2"x4" PLATE

2"x8" CLG JOISTS
@16" O.C.

(3) 2"x10" BEAM ON
PT 6"x6" POST

2"x4" FIRESTOP

2"x4" STUDS
@16" O.C.

8'-4½"

8'-6" CLEARANCE

PT 2"x4" SILL

12"Ø CONC
PIER ON FTG

BELOW
FROST LINE

Typical cross section A-A

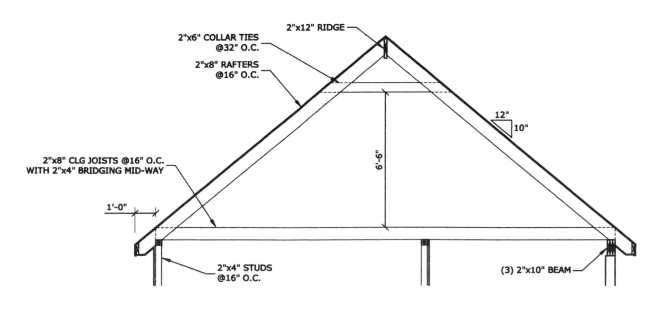

2"x6" COLLAR TIES
@32" O.C.

2"x12" RIDGE

2"x8" RAFTERS
@16" O.C.

12"
10"

2"x8" CLG JOISTS @16" O.C.
WITH 2"x4" BRIDGING MID-WAY

6'-6"

1'-0"

2"x4" STUDS
@16" O.C.

(3) 2"x10" BEAM

Alternate roof

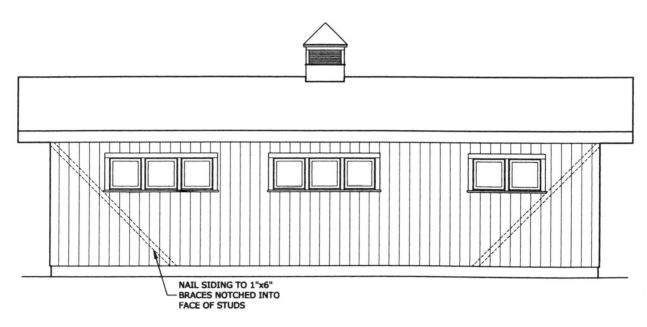

NAIL SIDING TO 1"x6"
BRACES NOTCHED INTO
FACE OF STUDS

Rear elevation

CLG = ceiling
CONC = concrete
Ø = diameter
FTG = footing
O.C. = on center
PT = pressure-treated
T&G = tongue and groove

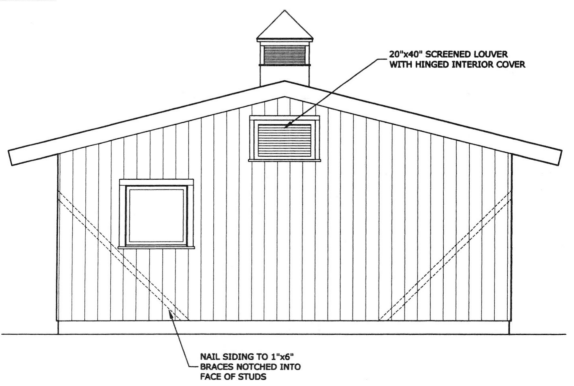

20"x40" SCREENED LOUVER
WITH HINGED INTERIOR COVER

NAIL SIDING TO 1"x6"
BRACES NOTCHED INTO
FACE OF STUDS

Right end elevation

Larger Stable

This design uses trusses instead of rafters
to support a double-wide span, and it accommodates
eight horses. Stall partitions are designed to prevent fighting
but are left open up high to improve ventilation. An alternative
to solid interior walls is a combination of wood and metal stall guards,
which provide visibility between all stalls while maintaining animal safety.
(See page 246 for plan credit.)

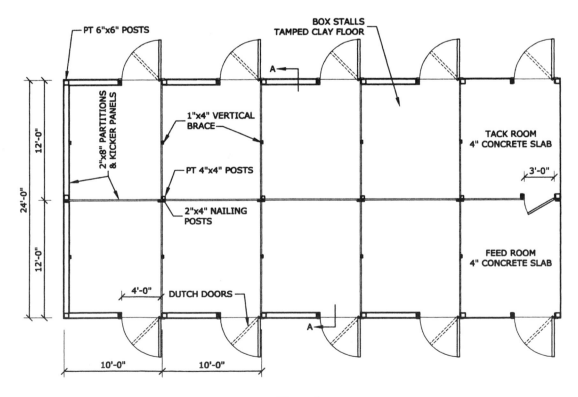

Floor plan

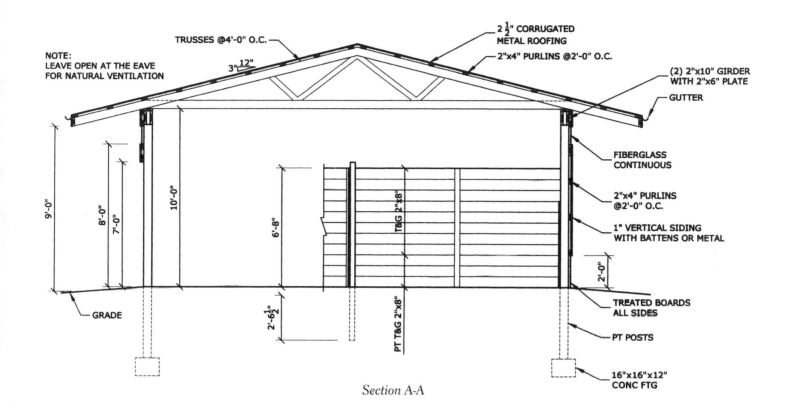

NOTE:
LEAVE OPEN AT THE EAVE
FOR NATURAL VENTILATION

TRUSSES @4'-0" O.C.

2 ½" CORRUGATED
METAL ROOFING

2"x4" PURLINS @2'-0" O.C.

3" 12

(2) 2"x10" GIRDER
WITH 2"x6" PLATE

GUTTER

FIBERGLASS
CONTINUOUS

2"x4" PURLINS
@2'-0" O.C.

1" VERTICAL SIDING
WITH BATTENS OR METAL

T&G 2"x8"

6'-8"

9'-0"

8'-0"

7'-0"

10'-0"

2'-0"

TREATED BOARDS
ALL SIDES

GRADE

2'-6½"

PT T&G 2"x8"

PT POSTS

16"x16"x12"
CONC FTG

Section A-A

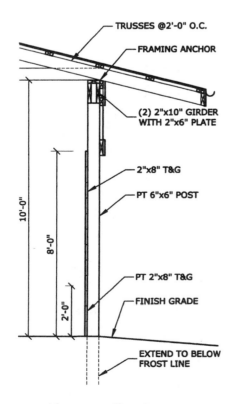

TRUSSES @2'-0" O.C.

FRAMING ANCHOR

(2) 2"x10" GIRDER
WITH 2"x6" PLATE

2"x8" T&G

PT 6"x6" POST

10'-0"

8'-0"

PT 2"x8" T&G

2'-0"

FINISH GRADE

EXTEND TO BELOW
FROST LINE

Alternate wall section

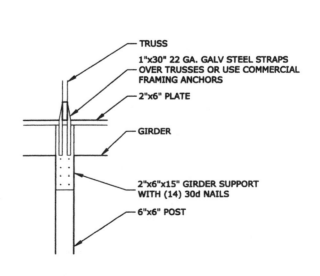

TRUSS

1"x30" 22 GA. GALV STEEL STRAPS
OVER TRUSSES OR USE COMMERCIAL
FRAMING ANCHORS

2"x6" PLATE

GIRDER

2"x6"x15" GIRDER SUPPORT
WITH (14) 30d NAILS

6"x6" POST

Post girder and nailing detail

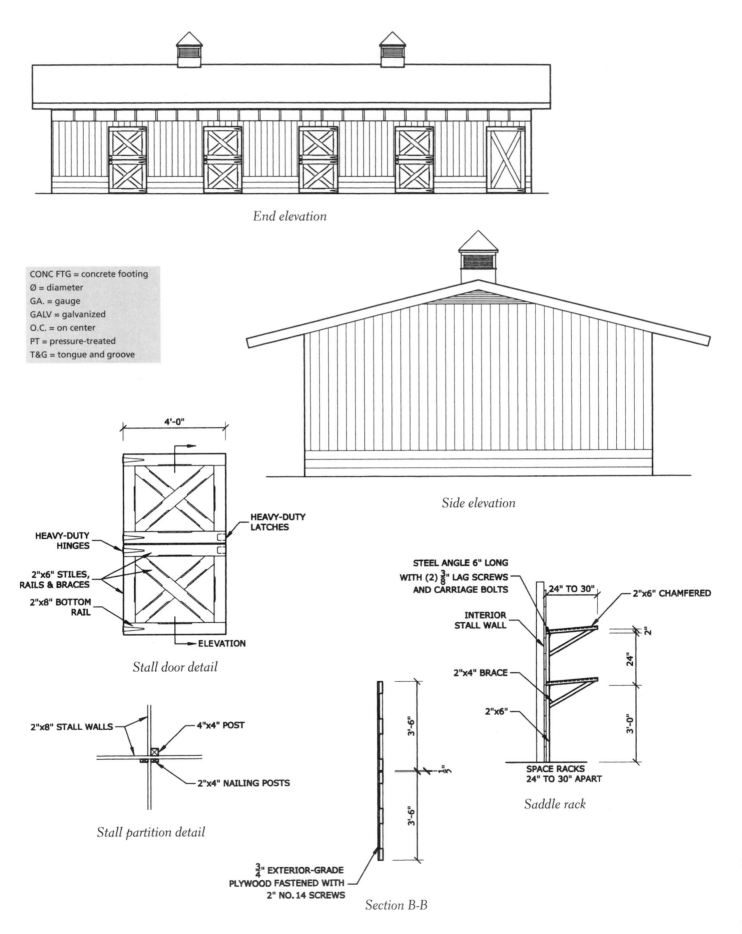

End elevation

CONC FTG = concrete footing
Ø = diameter
GA. = gauge
GALV = galvanized
O.C. = on center
PT = pressure-treated
T&G = tongue and groove

Side elevation

4'-0"

HEAVY-DUTY
LATCHES

HEAVY-DUTY
HINGES

2"x6" STILES,
RAILS & BRACES

2"x8" BOTTOM
RAIL

ELEVATION

Stall door detail

STEEL ANGLE 6" LONG
WITH (2) 3/8" LAG SCREWS
AND CARRIAGE BOLTS

24" TO 30"

2"x6" CHAMFERED

2"

INTERIOR
STALL WALL

24"

2"x4" BRACE

2"x6"

3'-0"

SPACE RACKS
24" TO 30" APART

Saddle rack

2"x8" STALL WALLS

4"x4" POST

2"x4" NAILING POSTS

Stall partition detail

3'-6"

1"

3'-6"

3/4" EXTERIOR-GRADE
PLYWOOD FASTENED WITH
2" NO. 14 SCREWS

Section B-B

Chicken Coop

Chickens are wonderful, productive animals and deserve a place on any farm; they even do well in backyards in town. From a small flock, you get meat, eggs, and hours of entertainment. A relatively small flock can also generate on-farm revenue. As consumers discover the flavor and health benefits of meat and eggs from birds raised under nonindustrial conditions, they seek out local sources of "farm-fresh" products.

Permanent chicken coops are usually divided in two with a screened door in between. The division between the two sides may be a simple chicken-wire barrier or, for larger structures, a center feed storage area with chicken-wire screening on either side of the storage area. The advantages of a divided coop include the ability to:

- raise young birds separate from adult hens on one side, as hens fuss at young birds and keep them from feed;

- thoroughly clean one side between rearings before moving birds to the other side and completely cleaning it;
- introduce new birds to the flock with less fighting (they get to know each other through the barrier before joining in one flock); and
- shuffle birds to one side or the other to catch an escapee.

Small coops usually have shed-style roofs, but gable-style roofs also work. The design shown here can easily be built on a unit-by-unit basis, with each unit providing 400 square feet of space. There are two units in this example, as shown in the plan, but extra units can be added to accommodate a larger operation. If hens are housed primarily inside, each unit can accommodate 100 hens (or 170 bantam hens); if hens have access to the outdoors, the units accommodate 50 percent more hens. (See page 246 for plan credit.)

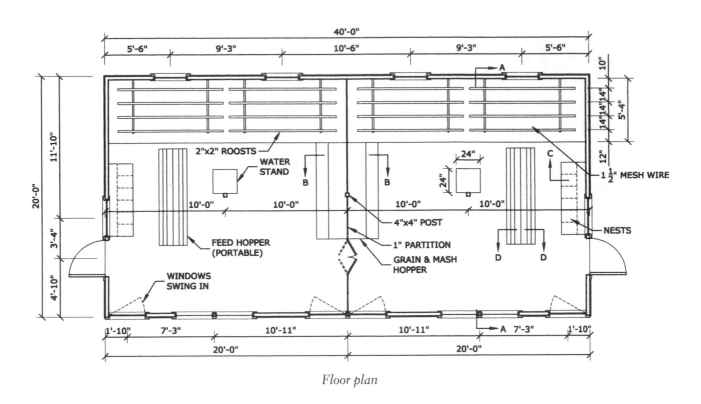

Floor plan

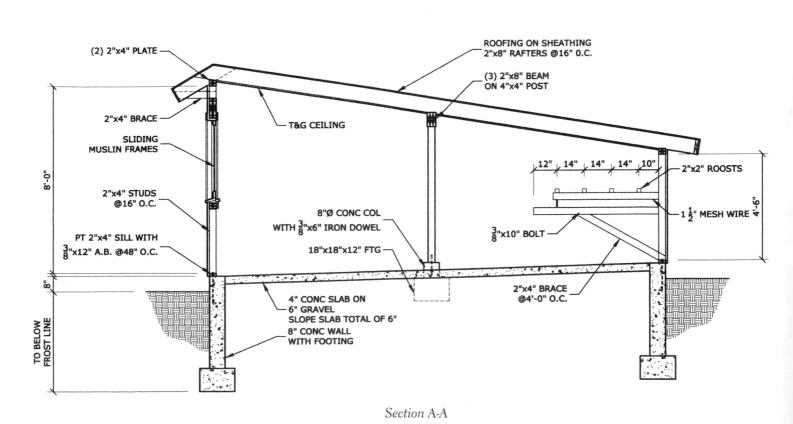

Section A-A

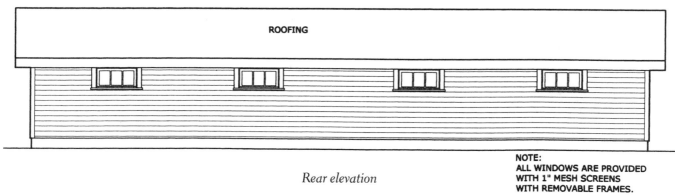

ROOFING

Rear elevation

NOTE:
ALL WINDOWS ARE PROVIDED
WITH 1" MESH SCREENS
WITH REMOVABLE FRAMES.

Side elevation

A.B. = anchor bolt
COL = column
CONC = concrete
Ø = diameter
FTG = footing
O.C. = on center
PT = pressure-treated
T&G = tongue and groove

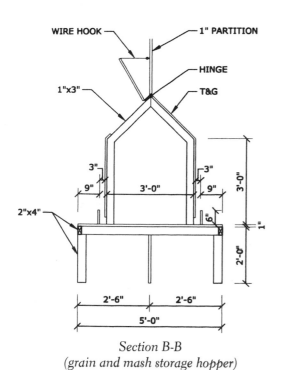

WIRE HOOK — 1" PARTITION
— HINGE
1"x3" — T&G
3" 3"
9" 9"
3'-0" 3'-0"
2"x4" 6"
1"
2'-0"
2'-6" 2'-6"
5'-0"

Section B-B
(grain and mash storage hopper)

SIDING — CASING
WINDOW SASH
SILL

Header

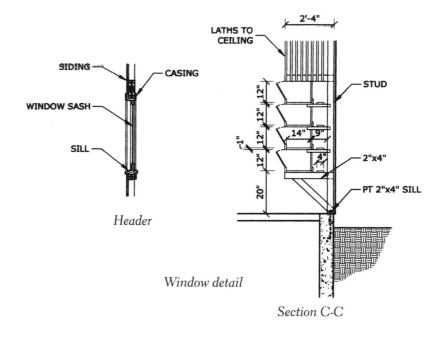

LATHS TO
CEILING
2'-4"
STUD
12" 12"
12" 12"
14" 9"
1"
12" 4"
2"x4"
20"
PT 2"x4" SILL

Window detail

Section C-C

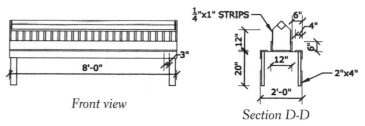

¼"x1" STRIPS
6"
4"
12"
3"
8'-0"
20"
12"
2'-0"
2"x4"

Front view

Section D-D

Portable mash hopper

LOAFING SHEDS

Loafing sheds are the most common farm structures, and 90 percent of them are built using post-barn techniques. There are myriad variations of these basic structures, as evidenced in the drawings at left and in the plans that follow. Your imagination, tempered by a little common sense, can conjure the perfect shed for your particular needs.

Although these sheds are usually finished in vertical steel (farmer's tin, or galvanized metal siding, is the most common, followed by colored metal panels), they can also be finished in lap siding (horizontal boards laid out so the bottom edge of each board overlaps the top of the board below it) or board-and-batten (vertical boards, with narrower boards, or battens, nailed over the seams). See page 215 for more information on siding types.

Loafing sheds sometimes have an enclosed area for feed, tack, tools, or equipment storage. They can be designed with the needs of certain species or age classes in mind, or they can be simple three-sided shelters any animal can use to get out of the weather. Partial closure on the front with a sliding door can turn an inexpensive shed into an enclosed barn.

Post structures can incorporate windows, entry doors, or large sliding doors. The detail drawings on the next several pages provide guidance on post-building construction techniques (see page 246 for plan credit), and they apply to just about any type of post building that you might be considering. Keep in mind that fully enclosed structures may require additional bracing for stability; three-sided post barns require additional gussets on the open edge.

Post-barn construction techniques are highly adaptable, and post structures are the most commonly built structures on farms today. These illustrations show several variations of the same loafing shed: the basic design (A), the basic design with two equal-width sections added to either side (B), and the basic design with an additional section added to one side only (C).

Walls, Doors, and Windows for Pole-type Buildings

Pole-type construction is common for barns and sheds. These plans illustrate basic design details for constructing wall, doors, and windows in a typical post-barn building. (See page 246 for plan credit.)

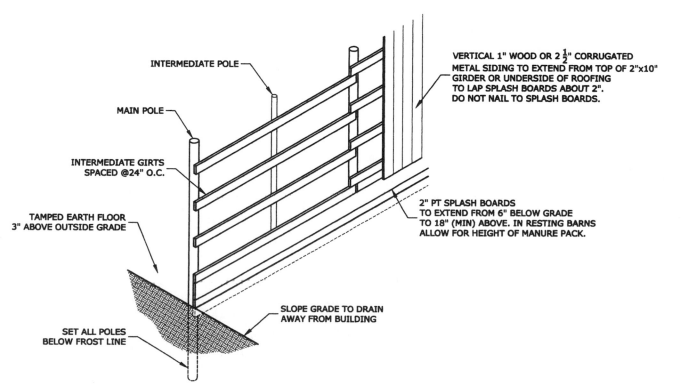

METAL FLASHING AT OPENINGS NOT PROTECTED BY ROOF OVERHANG

2"x8" CASING WITH 1"x3" STOP

2"x6" NOTCHED FOR GIRTS

2"x4" BLOCKING ON POST

1"x6" RAILS & BRACES

HINGE SIDE

VERTICAL 1" T&G

4'-0" MAX FOR SINGLE HINGED DOOR PANEL

NOTCH LOWER 2" SPLASH BOARDS TO FLOOR HEIGHT

HEIGHT TO SUIT

Typical framing for hinged door

INTERMEDIATE POLE

MAIN POLE

INTERMEDIATE GIRTS SPACED @24" O.C.

TAMPED EARTH FLOOR 3" ABOVE OUTSIDE GRADE

VERTICAL 1" WOOD OR 2½" CORRUGATED METAL SIDING TO EXTEND FROM TOP OF 2"x10" GIRDER OR UNDERSIDE OF ROOFING TO LAP SPLASH BOARDS ABOUT 2". DO NOT NAIL TO SPLASH BOARDS.

2" PT SPLASH BOARDS TO EXTEND FROM 6" BELOW GRADE TO 18" (MIN) ABOVE. IN RESTING BARNS ALLOW FOR HEIGHT OF MANURE PACK.

SLOPE GRADE TO DRAIN AWAY FROM BUILDING

SET ALL POLES BELOW FROST LINE

Typical wall construction (roof framing not shown)

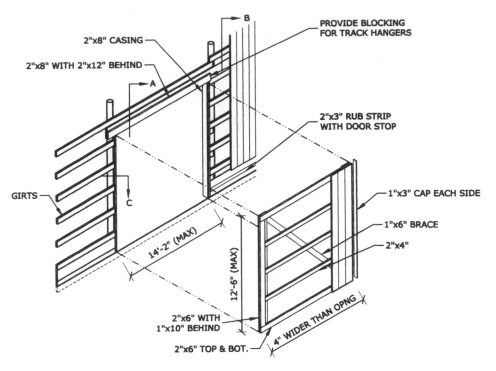

2"x8" CASING
2"x8" WITH 2"x12" BEHIND
GIRTS
14'-2" (MAX)
12'-6" (MAX)
2"x6" WITH 1"x10" BEHIND
2"x6" TOP & BOT.

PROVIDE BLOCKING FOR TRACK HANGERS
2"x3" RUB STRIP WITH DOOR STOP
1"x3" CAP EACH SIDE
1"x6" BRACE
2"x4"
4" WIDER THAN OPNG

Typical framing for sliding door

SIDING
2"x12"
2"x8"
2"x8" CASING
1"x2" FILLER
DOOR
FLASHING NOT REQUIRED UNDER EAVES
TRACK & (3) HANGERS INSTALLED AS RECOMMENDED BY MFR.
METAL CAP

Section A-A

SIDING
2"x12"
PT BLOCKING AT TRACK HANGERS
TRACK
DOOR

Section B-B

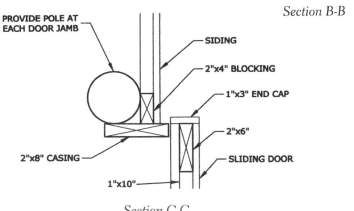

PROVIDE POLE AT EACH DOOR JAMB
SIDING
2"x4" BLOCKING
1"x3" END CAP
2"x6"
SLIDING DOOR
2"x8" CASING
1"x10"

Section C-C

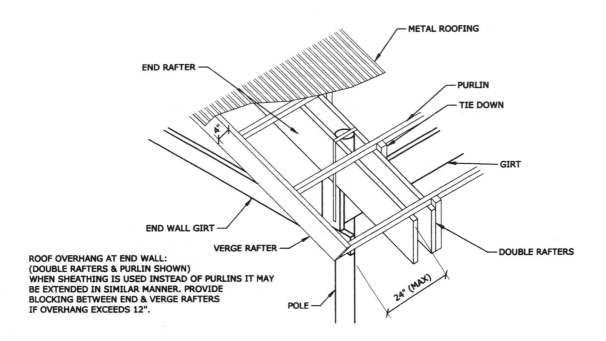

ROOF OVERHANG AT END WALL:
(DOUBLE RAFTERS & PURLIN SHOWN)
WHEN SHEATHING IS USED INSTEAD OF PURLINS IT MAY
BE EXTENDED IN SIMILAR MANNER. PROVIDE
BLOCKING BETWEEN END & VERGE RAFTERS
IF OVERHANG EXCEEDS 12".

Roof overhang at end wall

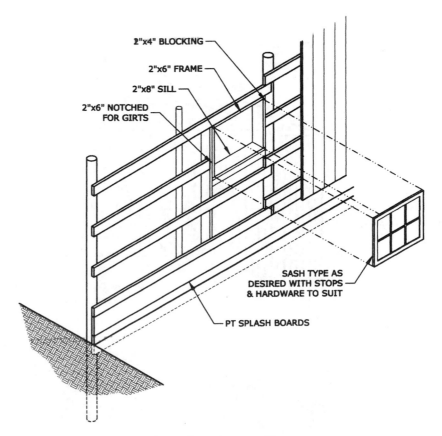

BOT = bottom
MAX = maximum
MIN = minimum
O.C. = on center
OPNG = opening
PT = pressure-treated
T&G = tongue and groove

Typical window installation

Basic Loafing Shed

This is a good, basic, three-sided loafing shed that animals will
get into by choice when the weather is bad or the sun is too hot. As shown, a little
fencing can readily be used to separate part of it for storage, but let me offer one bit of advice:
If you will store feed or anything that might be toxic to your animals, make the barrier between
the storage area and the critters a high, solid wall. (See page 246 for plan credit.)

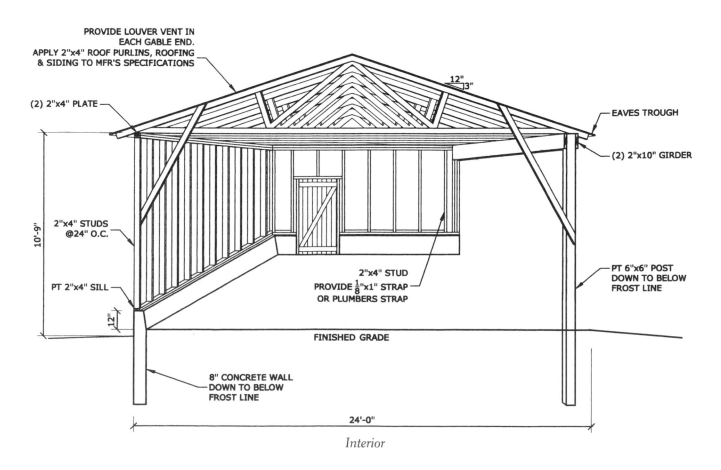

PROVIDE LOUVER VENT IN
EACH GABLE END.
APPLY 2"x4" ROOF PURLINS, ROOFING
& SIDING TO MFR'S SPECIFICATIONS

(2) 2"x4" PLATE

12"
3"

EAVES TROUGH

(2) 2"x10" GIRDER

2"x4" STUDS
@24" O.C.

10'-9"

PT 2"x4" SILL

12"

2"x4" STUD
PROVIDE $\frac{1}{8}$"x1" STRAP
OR PLUMBERS STRAP

PT 6"x6" POST
DOWN TO BELOW
FROST LINE

FINISHED GRADE

8" CONCRETE WALL
DOWN TO BELOW
FROST LINE

24'-0"

Interior

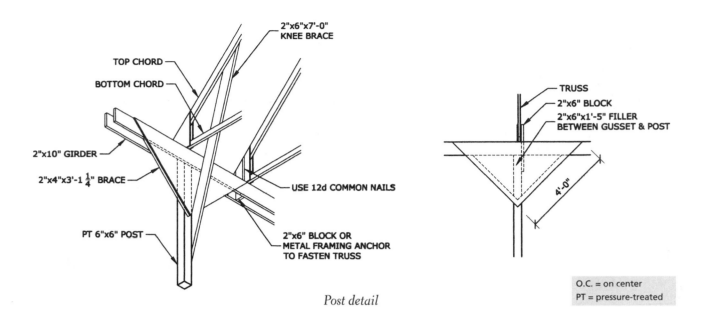

2"x6"x7'-0"
KNEE BRACE

TOP CHORD

BOTTOM CHORD

2"x10" GIRDER

2"x4"x3'-1¼" BRACE

USE 12d COMMON NAILS

PT 6"x6" POST

2"x6" BLOCK OR
METAL FRAMING ANCHOR
TO FASTEN TRUSS

Post detail

TRUSS

2"x6" BLOCK

2"x6"x1'-5" FILLER
BETWEEN GUSSET & POST

4'-0"

O.C. = on center
PT = pressure-treated

Sheep or Goat Shed

This shed is an excellent, all-purpose
shed for sheep or goats. It provides room
for feed storage, lambing/kidding pens, and a creep feed setup;
creep is a feed pen that only young animals can access, allowing them
to get extra feed while preventing mature animals from overfeeding. The ewes
or does and their offspring can feed from the hay bunk, which provides wind protection
for the pens in the rear corner. (See page 246 for plan credit.)

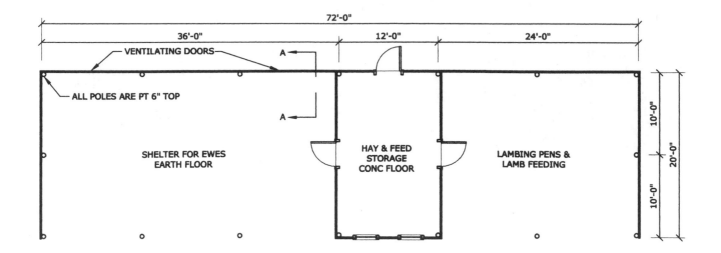

72'-0"

36'-0"   12'-0"   24'-0"

VENTILATING DOORS

A

ALL POLES ARE PT 6" TOP

A

SHELTER FOR EWES
EARTH FLOOR

HAY & FEED
STORAGE
CONC FLOOR

LAMBING PENS &
LAMB FEEDING

10'-0"

10'-0"

20'-0"

Floor plan

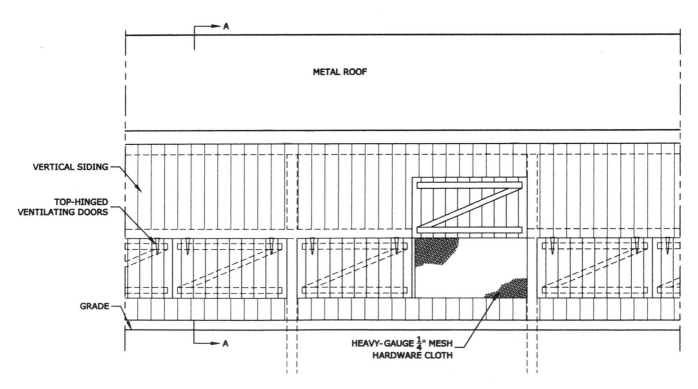

A

METAL ROOF

VERTICAL SIDING

TOP-HINGED
VENTILATING DOORS

GRADE

A

HEAVY-GAUGE $\frac{1}{4}$" MESH
HARDWARE CLOTH

Part rear elevation

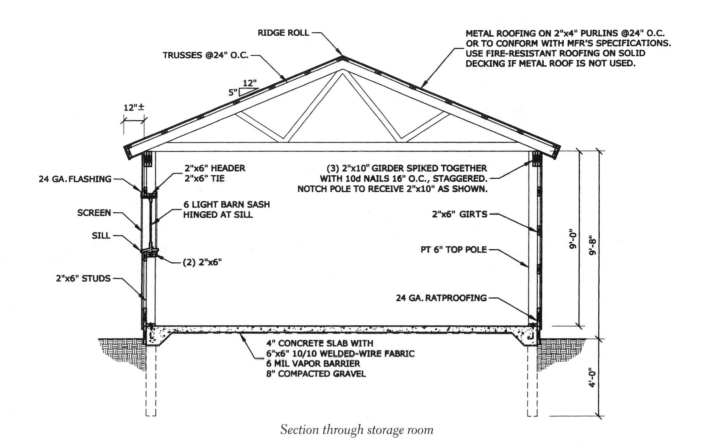

RIDGE ROLL

TRUSSES @24" O.C.

12"
5"

12"±

24 GA. FLASHING

2"x6" HEADER
2"x6" TIE

METAL ROOFING ON 2"x4" PURLINS @24" O.C.
OR TO CONFORM WITH MFR'S SPECIFICATIONS.
USE FIRE-RESISTANT ROOFING ON SOLID
DECKING IF METAL ROOF IS NOT USED.

(3) 2"x10" GIRDER SPIKED TOGETHER
WITH 10d NAILS 16" O.C., STAGGERED.
NOTCH POLE TO RECEIVE 2"x10" AS SHOWN.

SCREEN

6 LIGHT BARN SASH
HINGED AT SILL

2"x6" GIRTS

SILL

PT 6" TOP POLE

(2) 2"x6"

2"x6" STUDS

9'-0"
9'-8"

24 GA. RATPROOFING

4" CONCRETE SLAB WITH
6"x6" 10/10 WELDED-WIRE FABRIC
6 MIL VAPOR BARRIER
8" COMPACTED GRAVEL

4'-0"

Section through storage room

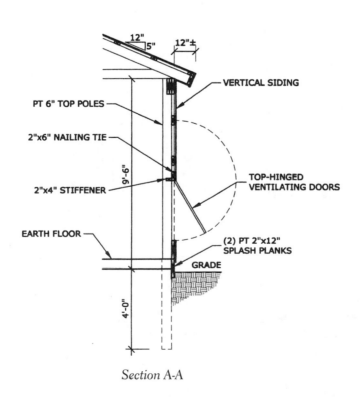

12"
5"

12"±

VERTICAL SIDING

PT 6" TOP POLES

2"x6" NAILING TIE

9'-6"

2"x4" STIFFENER

TOP-HINGED
VENTILATING DOORS

EARTH FLOOR

(2) PT 2"x12"
SPLASH PLANKS

GRADE

4'-0"

CONC = concrete
GA. = gauge
O.C. = on center
PT = pressure-treated

Section A-A

Enclosed Shed

This style of shed works well for all animals,
though the floor plan provided is intended as a sheep shed
in which lambs will be born in harsh weather. The doors can be closed
in extreme weather, creating an enclosed barn, but on nice days they can
be left open for ventilation and sunshine. (See page 246 for plan credit.)

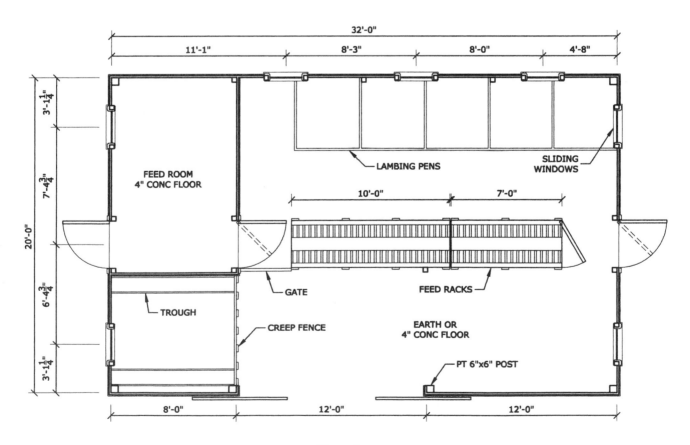

Floor plan

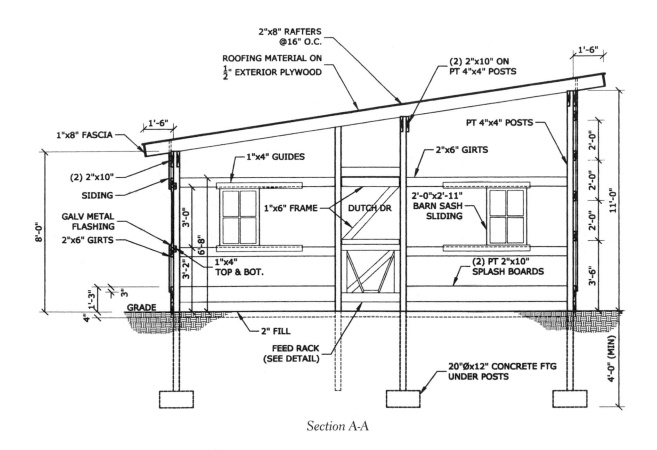

2"x8" RAFTERS @16" O.C.

ROOFING MATERIAL ON ½" EXTERIOR PLYWOOD

(2) 2"x10" ON PT 4"x4" POSTS

1'-6"

1'-6"

1"x8" FASCIA

(2) 2"x10"

SIDING

GALV METAL FLASHING

2"x6" GIRTS

8'-0"

1"x4" GUIDES

1"x6" FRAME

DUTCH DR

3'-0"

6'-8"

1"x4" TOP & BOT.

3'-2"

PT 4"x4" POSTS

2"x6" GIRTS

2'-0"x2'-11" BARN SASH SLIDING

(2) PT 2"x10" SPLASH BOARDS

2'-0"

2'-0"

2'-0"

3'-6"

11'-0"

GRADE

3"

1'-3"

4"

2" FILL

FEED RACK (SEE DETAIL)

20"Øx12" CONCRETE FTG UNDER POSTS

4'-0" (MIN)

Section A-A

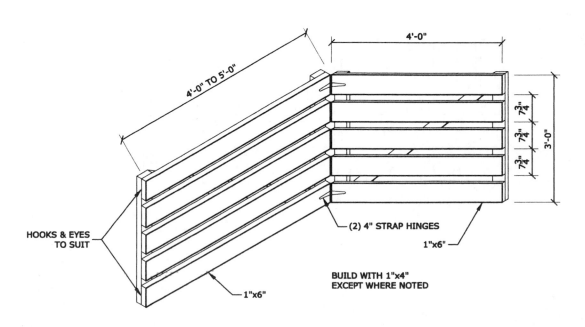

4'-0"

4'-0" TO 5'-0"

7¾"

7¾"

7¾"

3'-0"

HOOKS & EYES TO SUIT

(2) 4" STRAP HINGES

1"x6"

BUILD WITH 1"x4" EXCEPT WHERE NOTED

1"x6"

Hinged panels for temporary lambing pens

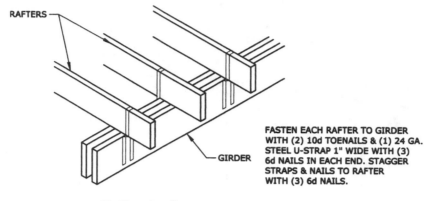

FASTEN EACH RAFTER TO GIRDER
WITH (2) 10d TOENAILS & (1) 24 GA.
STEEL U-STRAP 1" WIDE WITH (3)
6d NAILS IN EACH END. STAGGER
STRAPS & NAILS TO RAFTER
WITH (3) 6d NAILS.

Nailing detail

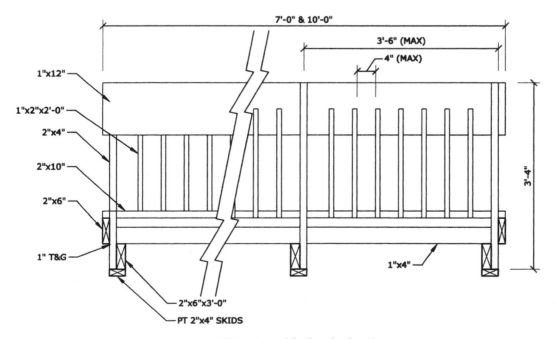

Elevation of feed racks detail

BOT = bottom
CONC = concrete
Ø = diameter
DR = door
FTG = footing
GA. = gauge
GALV = galvanized
MAX = maximum
MIN = minimum
O.C. = on center
PT = pressure-treated

Sheep Shed

This is a variation of the previous shed; it has open sides and no sliding doors. If you need to work animals in a small area, as at shearing time for sheep, use fencing or stock panels on the open sides to catch animals. (See page 246 for plan credit.)

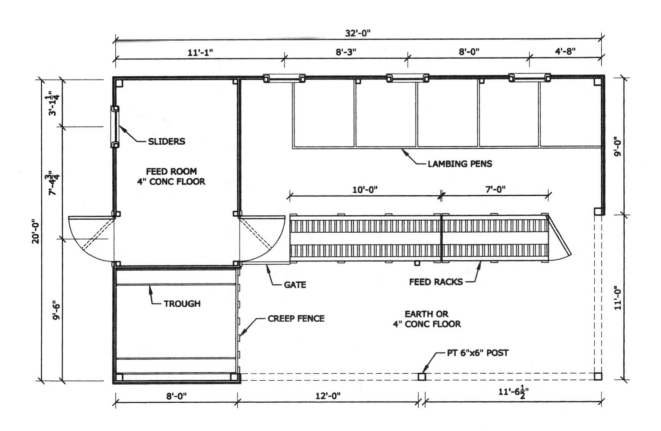

Floor plan

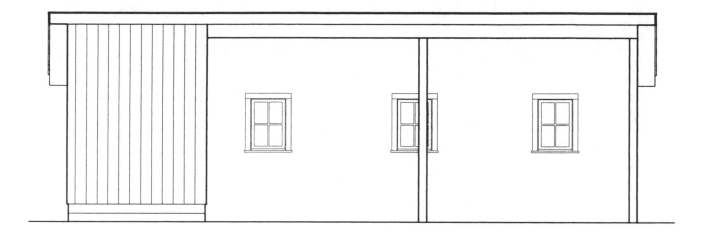

Front elevation

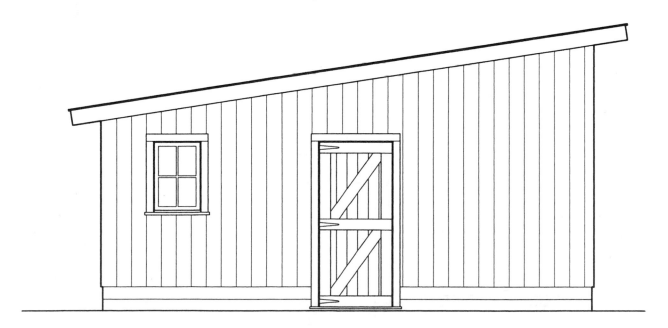

End elevation

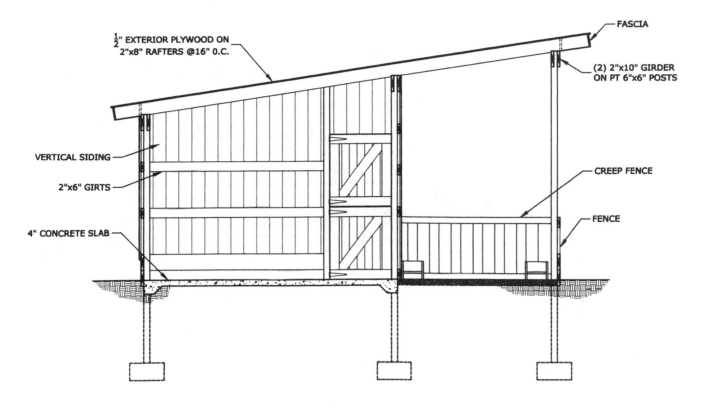

½" EXTERIOR PLYWOOD ON
2"x8" RAFTERS @16" O.C.

FASCIA

(2) 2"x10" GIRDER
ON PT 6"x6" POSTS

VERTICAL SIDING

2"x6" GIRTS

CREEP FENCE

FENCE

4" CONCRETE SLAB

Section A-A

CONC = concrete
O.C. = on center
PT = pressure-treated

Calf Shed

Lots of people who live in areas of the country where there's a significant number of dairy operations try their hand at raising dairy calves for replacement heifers (females that are raised to be sold back to a dairy farmer for use in the milking string) or meat. It seems like a great deal at first, because these calves are readily available and often quite cheap, but it can be a risky proposition: Dairy calves are taken away from their mothers almost immediately; good farmers make sure they get at least two days on their mom so that they get colostrum and a reasonable start, but some less scrupulous operators will sell their calves without this critical feed, sending them out into the world at a great disadvantage.

Even the calves that do get colostrum before being separated from their dams are vulnerable to diseases, so ventilation and cleanliness are even more critical for their survival than for any other class of livestock.

This structure is a cleverly designed shed that gives you a better chance of succeeding at a calf-rearing operation: fresh air is abundant and the sun warms the calves in winter but is blocked in summer. By using individual calf hutches within the structure, you reduce the risk of one calf spreading disease to its neighbor. The folks we know who have been successful at calf-rearing enterprises run a complete in-and-out system in which they buy a group of calves to rear and then, when the whole group is old enough and leaves the shed, thoroughly clean and disinfect the area before bringing in a new group. (See page 246 for plan credit.)

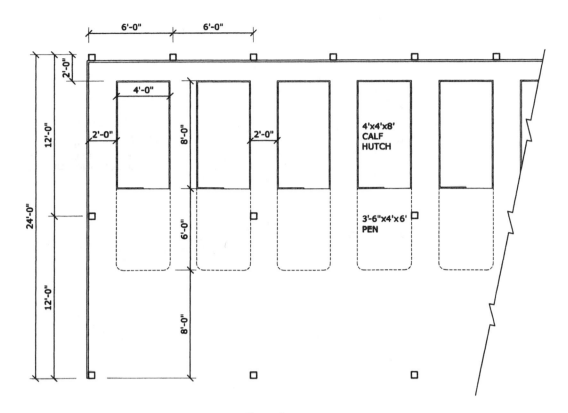

Floor plan

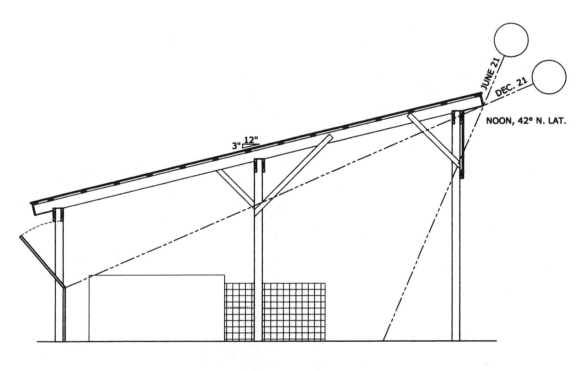

Solar penetration

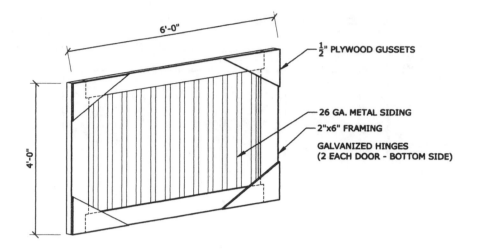

6'-0"

4'-0"

½" PLYWOOD GUSSETS

26 GA. METAL SIDING

2"x6" FRAMING

GALVANIZED HINGES
(2 EACH DOOR - BOTTOM SIDE)

Summer vent door

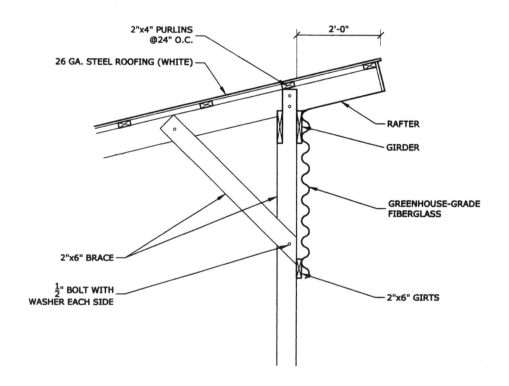

2"x4" PURLINS
@24" O.C.

26 GA. STEEL ROOFING (WHITE)

2'-0"

RAFTER

GIRDER

GREENHOUSE-GRADE
FIBERGLASS

2"x6" BRACE

½" BOLT WITH
WASHER EACH SIDE

2"x6" GIRTS

Front post detail, side view

GA. = gauge
O.C. = on center

What to Do about Pests

The first step in dealing with problem wildlife is to remove whatever is attracting it or build a barrier to what it finds attractive. For example, discourage swallows from nesting in your barn by swallow proofing before they get busy building their nests. Close off the space under eaves by stapling fine wire mesh (½ inch or smaller) at an angle. This approach also reduces woodpecker problems, as most flickers and other woodpeckers start drilling holes high on walls.

To keep bats and rodents from moving into the barn, screen louvers, vents, and fan openings; keep doors and windows in good repair; and use a chimney cap to keep critters from coming down the chimney. Rodents and small mammals such as rabbits are attracted to brush piles, logs, rock walls, and stones, and so a good way to keep them away is to remove these types of enticements.

Chicken-wire fencing buried to a depth of about 6 inches can keep rabbits, skunks, and other burrowing raiders out of gardens and orchards. Use buried chicken wire along the edges of chicken coops to keep out small mammals that feast on birds and eggs.

Install screens on all openings to minimize flies. Keep doors closed and windows shaded (used burlap feed sacks make fine window shades) to reduce fly pressure. Sawdust and wood shavings (from pine) are better bedding materials than straw because they contain a natural chemical compound that bugs don't like. Traps such as sticky tapes and jars with attractant are effective for the odd flies that get into a screened building.

Rats and mice generally come for stored feed. Storing grain and other feed away from the barn helps, but then you need to move it. Doors and windows that close tight will help keep them out, but storing food in varmint-proof containers is a key. On a small operation, heavy-plastic trash cans with tight-fitting lids work well for storing grain.

The second step to pest control is to attract beneficial creatures to your farm, notably bats, barn owls, purple martins, and bluebirds, among others. Bats are important members of the ecosystem. The strategic placement of bat houses can attract these worthy, unfairly maligned creatures to your place. You don't want them living inside the barn, but if they live on your farm, you will benefit from their voracious appetite; they can put a significant dent in the insect population. Owls do a great job of controlling rodent populations that eat crops and stored feed, and are relatively easy to attract to owl houses. Bluebirds are excellent insect eaters. Their numbers were plummeting for years, but bluebird houses and "bluebird trails" have helped them make a comeback over much of their range. The largest American swallows, Purple Martins are gregarious, colony-living birds. They feed on mosquitoes, wasps, moths, flies, grasshoppers, bees, ballooning spiders, midges, dragonflies, damselflies, cicadas, stinkbugs, beetles, and butterflies. Providing desirable housing and a welcome habitat for these beneficial creatures will keep them working for you year after year.

Know the Law

All wild birds — except pigeons, English Sparrows, and European Starlings — are protected by federal and state laws: you may *not* trap, kill, or possess protected species without federal and state permits. Most other wildlife is also protected by a variety of laws and regulations. Check with the U.S. Fish & Wildlife Service office in your region, your state's game and fish department or division of wildlife, or your Extension agent. These folks can suggest specific steps for dealing with problems, and in certain cases, like problem bears, they may trap and remove the critter(s) that are causing problems.

7. ODDS & ENDS

When you have animals, all kinds of appurtenances are either absolutely necessary or come in handy; they make your life easier and the lives of your animals healthier and safer. Ken and I tend not to be big shoppers, and over the years this has sometimes hurt us: we have decided not to purchase something because we thought it was too expensive, only to realize later that we were penny-wise and pound-foolish. This chapter provides advice on items to purchase and those to build.

CARTS AND CARRIERS

A two-wheeled utility cart and/or a really good, two-wheel wheelbarrow with a plastic tub are essential farmstead tools. Whether you are clearing out and moving manure, hauling feed, carting transplants, or carrying tools, you'll use them almost every day.

As for motorized wheels, larger operations can justify a tractor, but what about the small-scale hobbyist? Consider buying an all-terrain vehicle; used ATVs are readily available. They are great to have, and these days you can get a

wide variety of implements specially designed for use with them. We have a super little manure spreader (from Newer Spreader; see resources), seeder, and utility trailer for ours, and these tools have reduced the workload around here considerably.

STOCK PANELS

For years we didn't purchase stock panels because they seemed like an extravagance we could live without; the first couple we did buy we got used at a farm auction. These initial units were rusty and beat up but the price was right, and from those few old beaters we quickly realized what we had been missing. Now we have a dozen, and we wouldn't think of life without panels: they are incredibly versatile. Configure them into a round pen for training horses; place them around haystacks to prevent unauthorized eating; use them to round up sick animals for treatment; create lanes with them to funnel animals into a building; throw a tarp over them or stack straw bales around them to create a temporary shelter; wire boards (or tree limbs) to them to create a temporary windbreak; or place boards on top to make scaffolding for painting and construction projects. Stock panels can be purchased from most farm supply and feed stores.

For just a few small animals, you can build wooden panels or wood and welded-wire panels that will serve the purpose. Panels can be created as individual units or, for convenience, connected with hinges.

STOCK PANELS

Stock panels are indispensable when you are raising large animals. Metal stock panels can be purchased at farm supply stores and found used at auction. They are convenient but somewhat expensive when purchased new. Despite the cost, we prefer this type.

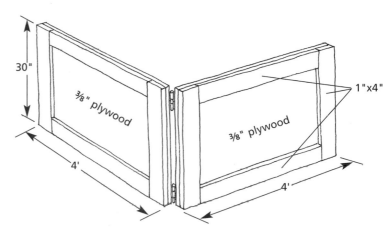

Plywood and 1x4 panels are easy to construct and are especially good for pigs. Because plywood is expensive, these aren't the bargains they once were, but constructing panels is an excellent way to use up plywood left over from other projects.

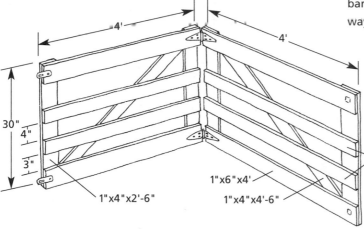

Native lumber is a good, economical choice for use in homemade panels. This design is suitable for sheep but not pigs; pigs will quickly destroy this type of panel. If you'll use the panel for cattle and horses, leave out the first 1x4 up from the base.

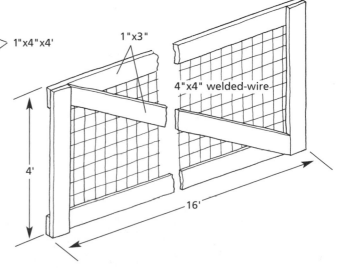

This type of wire mesh panel is fairly inexpensive to construct and is suitable for all classes of livestock, including poultry and fowl. Be sure animals are not exposed to sharp wire edges.

ITEMS FOR POULTRY AND SMALL ANIMAL CARE

Feeders and waterers for poultry and fowl are readily available in plastic and lightweight metal. They are generally cheap and easy to clean, so there's little reason to construct these items, but the large feeder shown here is convenient if you have a dozen or more birds and don't want to be continually filling feeders.

Waterers are best placed on a screened stand. This helps keep litter and manure out of the water so it stays fresh and clean.

Plastic feeder

Plastic feeders made with 5-gallon buckets are commercially available, economical, and hold up to 50 pounds of feed. To keep the feed clean, suspend the feeder from a rope so it's about an inch off the ground.

Wooden feeder

⅜"x1½"
1"x6"
1"x6"
1"x6"
11½"
1"x6"
5'
14"

Construct this feeder with scrap lumber from your chicken coop. The spinning board on top keeps birds from climbing in the feed, and because the feed is off the ground, it stays cleaner.

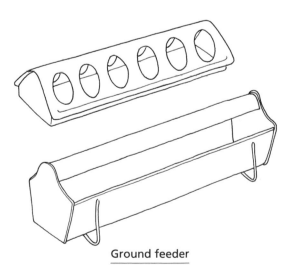

Ground feeder

These small feeders are good for chicks (the design of the top keeps them from walking in the feed) or for ground feed with a small flock. Heavy plastic and metal versions are available at farm supply stores; metal costs a bit more but is more durable than plastic.

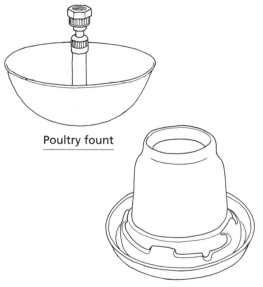

Poultry fount

Plastic jug waterer

Because it's automatic, a poultry fount is convenient for a large operation (100 birds or more), but it can freeze in winter. Plastic jug waterers are good for starting chicks and for small flocks.

Nest boxes are essential if you plan to keep layers and are expensive if you buy the commercially available type. You can make them out of wood or, for a few birds, plastic storage boxes (preferably in a solid color) by cutting a hole in the lid. For ducks, geese, or turkeys, an old wooden crate or wooden box on its edge works well — it needs to be at least 18 inches by 24 inches, but bigger is fine. (Rabbits also need nest boxes.) Nest boxes should be lined with straw or shredded newspaper.

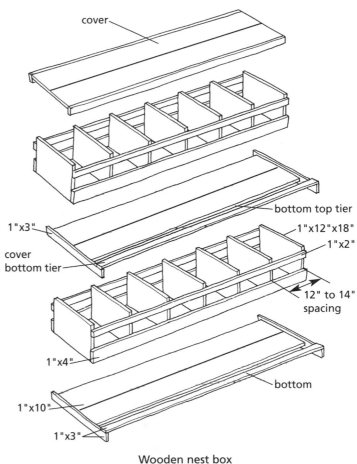

cover

bottom top tier

1"x3"

cover
bottom tier

1"x12"x18"

1"x2"

12" to 14"
spacing

1"x4"

bottom

1"x10"

1"x3"

Wooden nest box

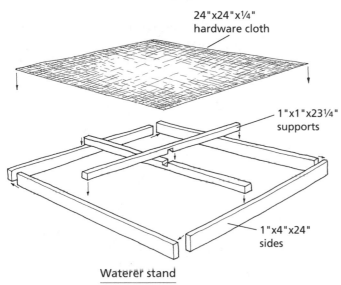

24"x24"x¼"
hardware cloth

1"x1"x23¼"
supports

1"x4"x24"
sides

Waterer stand

This stand is simple to construct. Apply wood glue before nailing the frame together. After the glue has set, staple hardware cloth to the frame. Placing waterers on the stand keeps water free of litter and dung.

This nest box can be built as a single- or two-level unit. Plan on one nest for every four birds if you won't let them brood eggs, or one nest for every two birds if you'll let broody hens sit on their clutch. Add 12 to 14 inches for each additional nest you need. (See page 247 for plan credit.)

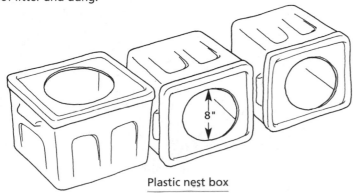

8"

Plastic nest box

Solid-colored plastic storage boxes can be used to make nest boxes and are helpful when you have a broody hen in a big flock. After dark, go into the henhouse and move the broody hen, eggs and all, into the box. Set it away from the other nest boxes so the other chickens won't disturb her or steal her nest.

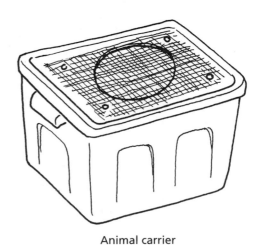

Animal carrier

Plastic boxes also make handy animal carriers for poultry and other small animals. (See page 139 for instructions.)

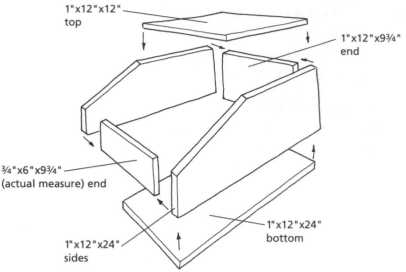

1"x12"x12" top

1"x12"x9¾" end

¾"x6"x9¾" (actual measure) end

1"x12"x24" sides

1"x12"x24" bottom

Nest box

Rabbits need nest boxes if you plan to breed them. You can purchase metal rabbit nest boxes from farm supply stores or build this wooden model. Clean out nest boxes well between litters.

This carrying cage is practical for transporting any small animals and can be constructed in an afternoon. A cookie sheet slides into grooves cut in the bottom boards to catch droppings.

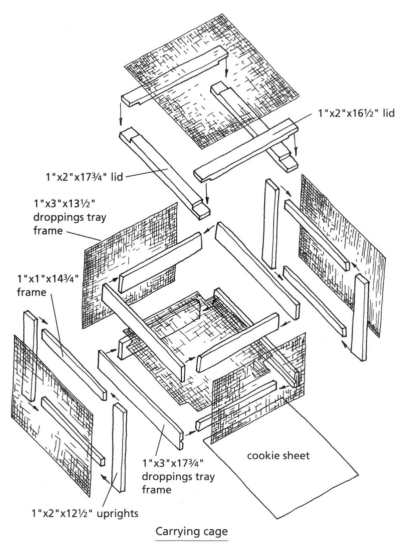

1"x2"x16½" lid

1"x2"x17¾" lid

1"x3"x13½" droppings tray frame

1"x1"x14¾" frame

1"x3"x17¾" droppings tray frame

cookie sheet

1"x2"x12½" uprights

Carrying cage

Birds like to roost up off the ground. Most domestic birds aren't great flyers anymore — they have been bred to have a large breast, which hinders their flying ability. Roosts should be placed at least 12 inches off the floor and they can be set in ladder fashion, with at least 8 inches between the rungs.

Small boxes for carrying birds or other small animals can be made of wood and chicken wire, or again you can use the plastic storage boxes with ¼-inch hardware cloth bolted into the hole cut in the top, with ¼-inch bolts set through holes drilled before the hole is cut in the lid. Also, drill four ¼-inch ventilation holes in two sides of the box and four ¼-inch drainage holes in the bottom of the box.

This roost has no floor or back; push it up against a wall in the barn or coop and pull it out occasionally to clean underneath. Solid sides and screening keep the birds out from under the roost.

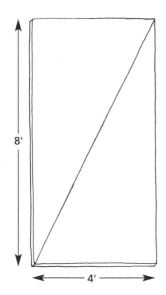

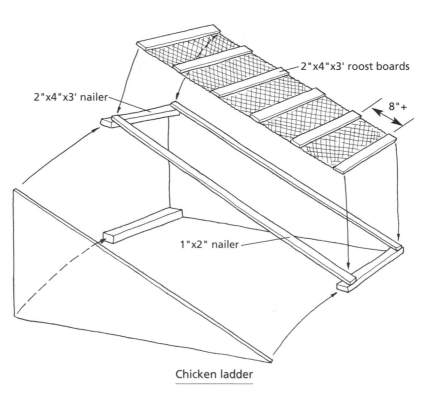

Chicken ladder

EQUIPMENT FOR SHEEP AND GOATS

Occasionally you will need to control a doe or ewe in a stanchion for medical treatment or to get her to adopt an orphan. For most small herds, one stanchion box is sufficient.

There are many styles of feed bunks for sheep and goats; I've included a few designs here. Lambs and kids need access to some extra ration, but ewes and does get piggy when given the opportunity and knock the kids out of the way to enjoy extra feed. The way to deal with this is to provide a "creep" feeder that young animals can access but not moms. This creep feeder design works well on pasture for a small group of babies but benefits from setting and wiring a T-post in the center of the wire cage to keep it in place and reduce the chance of the does or ewes bashing in the wire or climbing on it. The T-post doesn't have to be driven into the ground deeply; just set it so the bottom plate is buried about halfway, which will make it easy for you to pull it out and move the creep.

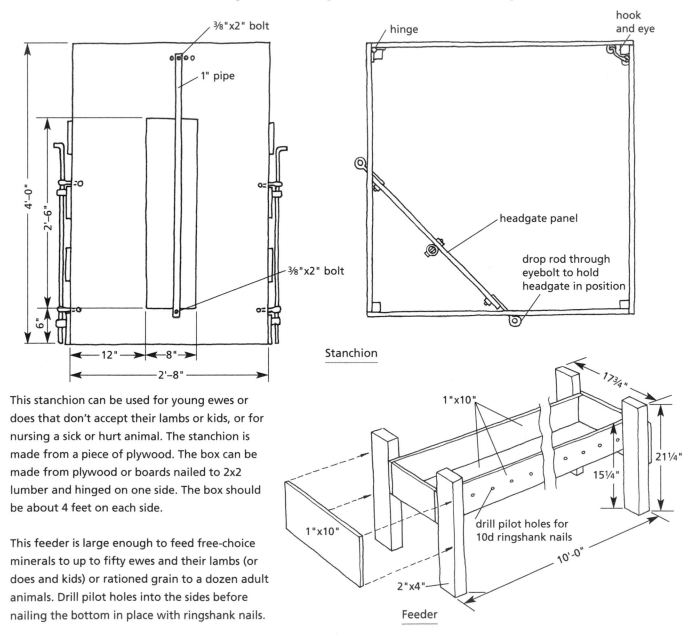

Stanchion

Feeder

This stanchion can be used for young ewes or does that don't accept their lambs or kids, or for nursing a sick or hurt animal. The stanchion is made from a piece of plywood. The box can be made from plywood or boards nailed to 2x2 lumber and hinged on one side. The box should be about 4 feet on each side.

This feeder is large enough to feed free-choice minerals to up to fifty ewes and their lambs (or does and kids) or rationed grain to a dozen adult animals. Drill pilot holes into the sides before nailing the bottom in place with ringshank nails.

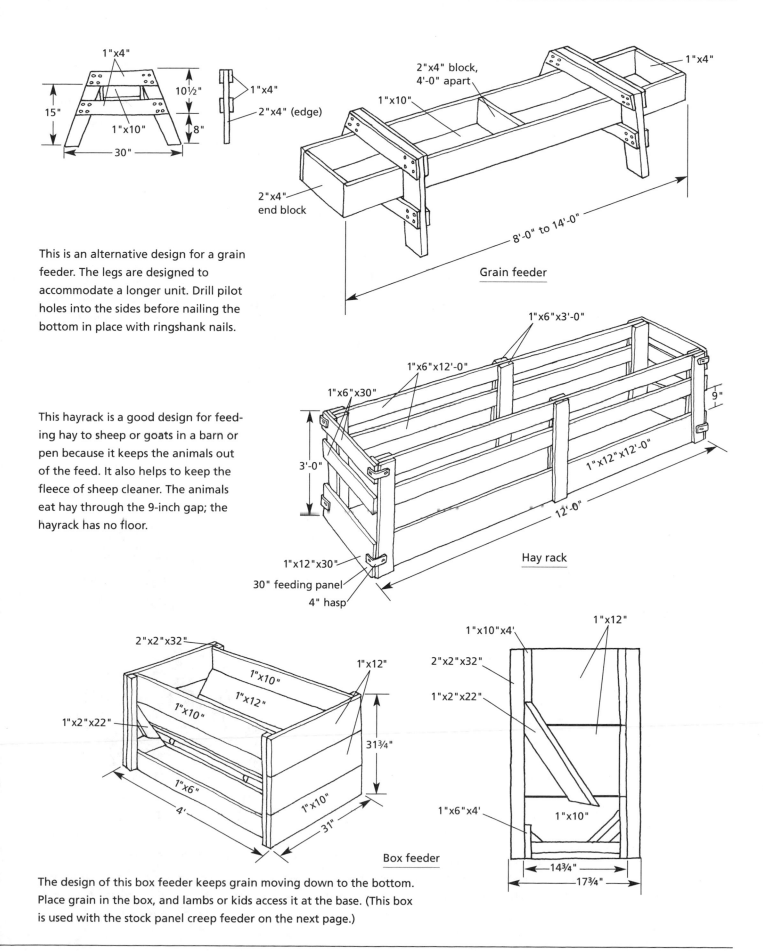

1"x4"

10½"

15"

1"x10"

30"

8"

1"x4"

2"x4" (edge)

2"x4" block, 4'-0" apart

1"x10"

1"x4"

2"x4" end block

8'-0" to 14'-0"

Grain feeder

This is an alternative design for a grain feeder. The legs are designed to accommodate a longer unit. Drill pilot holes into the sides before nailing the bottom in place with ringshank nails.

1"x6"x3'-0"

1"x6"x12'-0"

1"x6"x30"

3'-0"

9"

1"x12"x12'-0"

12'-0"

This hayrack is a good design for feeding hay to sheep or goats in a barn or pen because it keeps the animals out of the feed. It also helps to keep the fleece of sheep cleaner. The animals eat hay through the 9-inch gap; the hayrack has no floor.

1"x12"x30"

30" feeding panel

4" hasp

Hay rack

2"x2"x32"

1"x10"

1"x12"

1"x10"

1"x2"x22"

1"x6"

4'

31"

1"x10"

1"x12"

31¾"

Box feeder

1"x10"x4'

1"x12"

2"x2"x32"

1"x2"x22"

1"x6"x4'

1"x10"

14¾"

17¾"

The design of this box feeder keeps grain moving down to the bottom. Place grain in the box, and lambs or kids access it at the base. (This box is used with the stock panel creep feeder on the next page.)

When you are providing water in buckets in a barn, it tends to get fouled easily, or the animals dump the bucket, resulting in damp floors and inadequate water supplies. One way to minimize water problems is to wire the water bucket to a piece of plywood that has an 8-inch by 10-inch hole for access.

Sheep and goats don't need supplemental heat, but if you will be lambing or kidding in the winter, you will need to be able to provide supplemental heat to the babies for the first day or two after birth. A brooder warmed with lightbulbs does the trick.

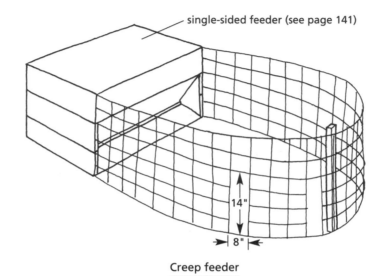

single-sided feeder (see page 141)

14"

8"

Creep feeder

A creep feeder allows young animals access to supplemental feed while preventing adults from stuffing themselves. This stock panel version for lambs and kids is excellent.

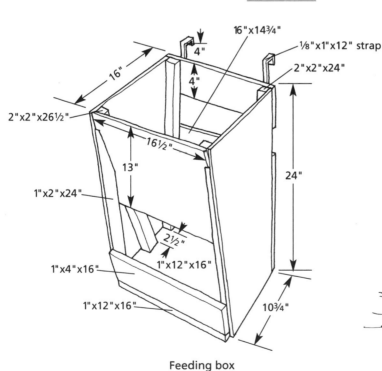

16"x14¾"

⅛"x1"x12" strap

2"x2"x24"

4"

16"

4"

2"x2"x26½"

16½"

13"

24"

1"x2"x24"

2½"

1"x4"x16"

1"x12"x16"

1"x12"x16"

10¾"

Feeding box

When animals are kept in small pens, hang a feeding box like this on the wall or over the fence and stuff it with hay. Because it's elevated, it keeps the hay clean.

Keyhole waterer

When watering one or two animals in a pen, this keyhole waterer helps keep water clean and in the bucket where it belongs.

Lamb/Kid Brooder

If just one or two lambs or kids are born in winter, you can bring them into the house to warm them up if they are cold, but for a larger winter lambing/kidding operation, a brooder is a valuable device. Keep a close eye on babies that are in a brooder; they can overheat quickly and may not be smart enough to move away from the light. (See page 247 for plan credit.)

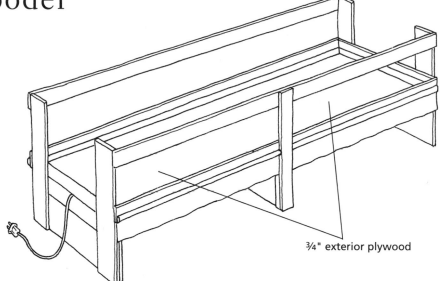

¾" exterior plywood

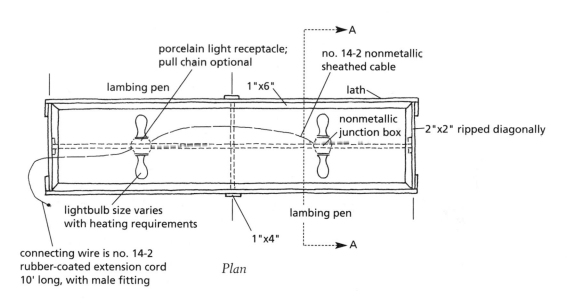

porcelain light receptacle; pull chain optional

no. 14-2 nonmetallic sheathed cable

lambing pen

1"x6"

lath

nonmetallic junction box

2"x2" ripped diagonally

lightbulb size varies with heating requirements

lambing pen

1"x4"

connecting wire is no. 14-2 rubber-coated extension cord 10' long, with male fitting

Plan

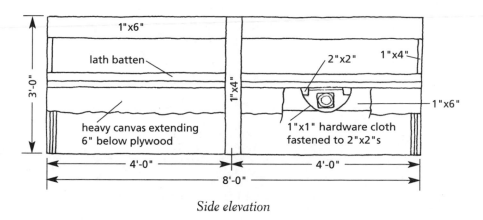

1"x6"

lath batten

2"x2"

1"x4"

3'-0"

1"x4"

1"x6"

heavy canvas extending 6" below plywood

1"x1" hardware cloth fastened to 2"x2"s

4'-0"

4'-0"

8'-0"

Side elevation

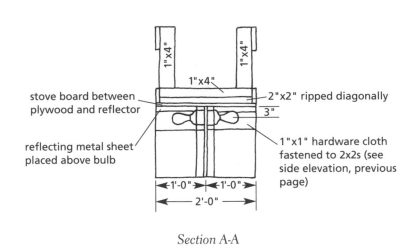

1"x4"   1"x4"

stove board between plywood and reflector

1"x4"

2"x2" ripped diagonally

3"

reflecting metal sheet placed above bulb

1"x1" hardware cloth fastened to 2x2s (see side elevation, previous page)

◄1'-0"►◄1'-0"►
◄——2'-0"——►

Section A-A

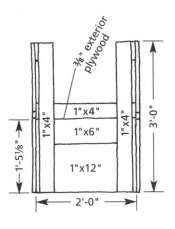

3/8" exterior plywood

1"x4"   1"x4"   1"x4"

1"x6"

1"x12"

1'-5⅛"

3'-0"

◄——2'-0"——►

End elevation

Keyhole stanchion

Make a keyhole stanchion if you need to milk only one or two animals. Feeding the animals a treat encourages them to place their heads through the hole.

Goat milk is excellent and good for your health. People who are allergic to cow's milk can often drink goat's milk with no problem, and it has smaller fat particles that are easier to digest. If you simply want to milk one goat for home use, a keyhole stanchion will work well (a chain at the top of the keyhole will keep the goat from raising her head while you are milking), but if you will be milking more than one goat, construct a stanchion platform to save your back from many aches. With either option, a milk stool will make life easier.

Because of concerns about fat intake and cholesterol levels and the increased availability of artisanal cheeses, there is a growing interest in both goat-milk and sheep-milk cheeses and other dairy products, making small commercial ventures viable. We know some people who, milking just a dozen goats only a few months a year, have a successful small business making goat-milk soaps and selling them at craft fairs and through local gift shops. For those considering a commercial venture, a small milking barn is essential. Sanitation regulations for commercial dairies are strict, so for ease of cleaning, stanchions for a commercial operation must be all metal instead of wood.

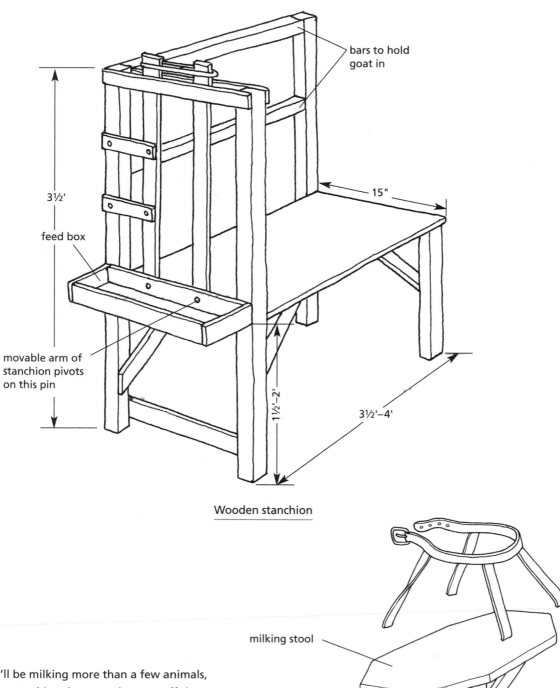

bars to hold
goat in

3½'

feed box

movable arm of
stanchion pivots
on this pin

15"

1½'–2'

3½'–4'

Wooden stanchion

milking stool

12"

If you'll be milking more than a few animals,
build a stanchion that gets them up off the
ground to minimize the strain on your back.
And be sure to use a milking stool. If you'll be
milking a small herd, you don't need the straps,
but for milking a larger herd, the straps are a
helpful convenience, keeping your hands free
while you work.

Milking Barn

Small-scale dairy operations need a barn that meets federal specifications. This plan shows just such a barn for a small goat (or sheep) dairy. The requirements are similar for a cow dairy, but the milking parlor would have to be larger. (See page 247 for plan credit.)

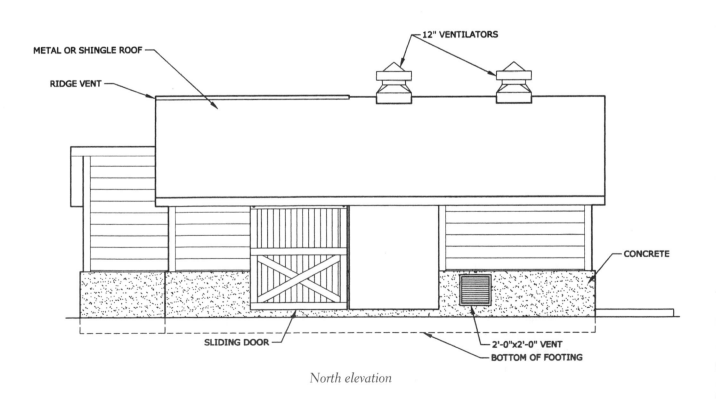

North elevation

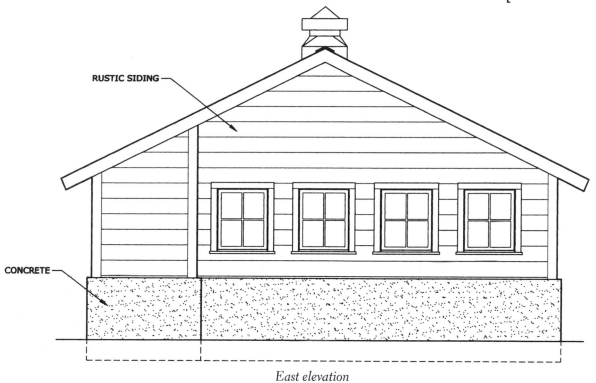

RUSTIC SIDING

CONCRETE

East elevation

COMP. = compressor
RM = room

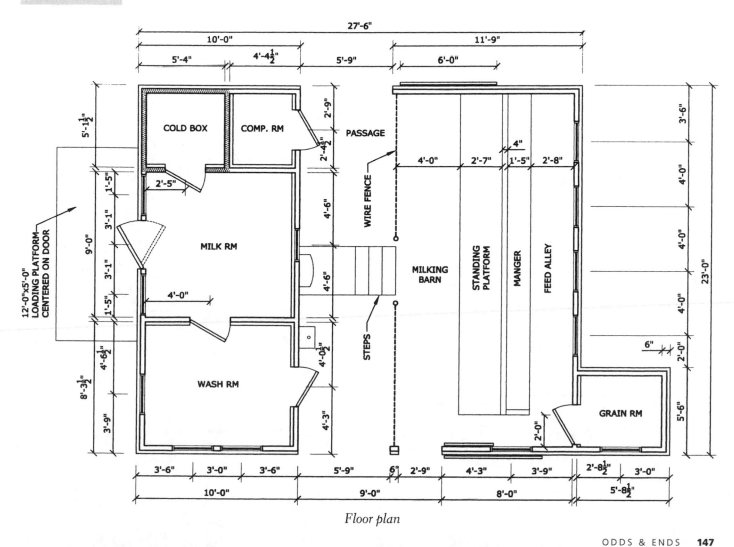

27'-6"

10'-0"
11'-9"

5'-4"
4'-4½"
5'-9"
6'-0"

5'-1½"

COLD BOX
COMP. RM
2'-9"
PASSAGE

2'-4½"

2'-5"
1'-5"
WIRE FENCE

4'-0"
2'-7"
1'-5"
2'-8"
4"

9'-0"
3'-1"

MILK RM
4'-6"

12'-0"x5'-0"
LOADING PLATFORM
CENTERED ON DOOR

3'-1"
4'-0"

MILKING
BARN

STANDING PLATFORM

MANGER

FEED ALLEY

3'-6"

4'-0"

4'-0"

1'-5"
4'-6"

STEPS

23'-0"

4'-6½"
8'-3½"

4'-0½"

4'-0"

6"
2'-0"

WASH RM
GRAIN RM

3'-9"
4'-3"
2'-0"

5'-6"

3'-6"
3'-0"
3'-6"
5'-9"
6"
2'-9"
4'-3"
3'-9"
2'-8½"
3'-0"

10'-0"
9'-0"
8'-0"
5'-8½"

Floor plan

Metal Stanchion

Because this stanchion is made of metal, it is suitable for use in a commercial dairy operation. Any professional welder will easily be able to construct it for you. (See page 247 for plan credit.)

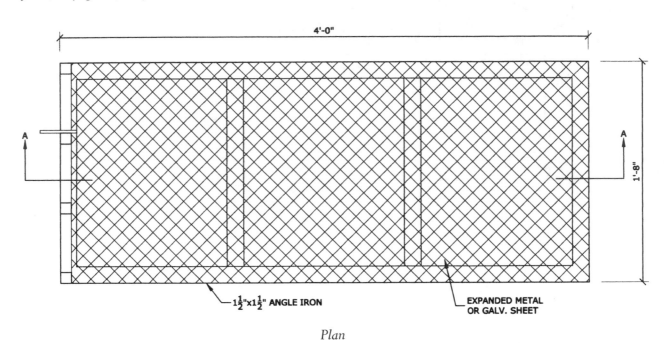

4'-0"

1'-8"

A

A

1½"x1½" ANGLE IRON

EXPANDED METAL OR GALV. SHEET

Plan

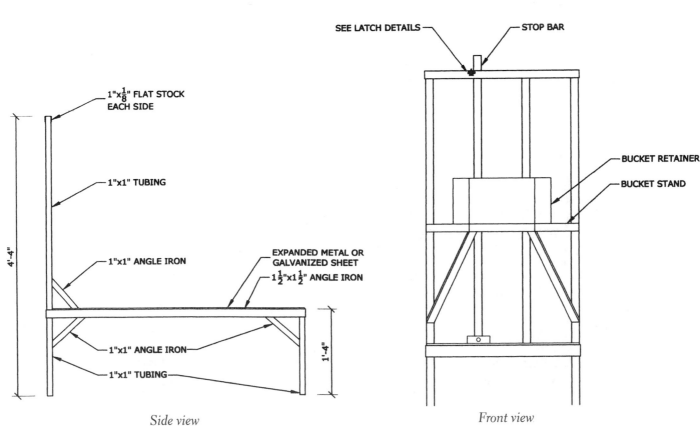

1"x⅛" FLAT STOCK EACH SIDE

1"x1" TUBING

1"x1" ANGLE IRON

4'-4"

EXPANDED METAL OR GALVANIZED SHEET

1½"x1½" ANGLE IRON

1"x1" ANGLE IRON

1"x1" TUBING

1'-4"

Side view

SEE LATCH DETAILS

STOP BAR

BUCKET RETAINER

BUCKET STAND

Front view

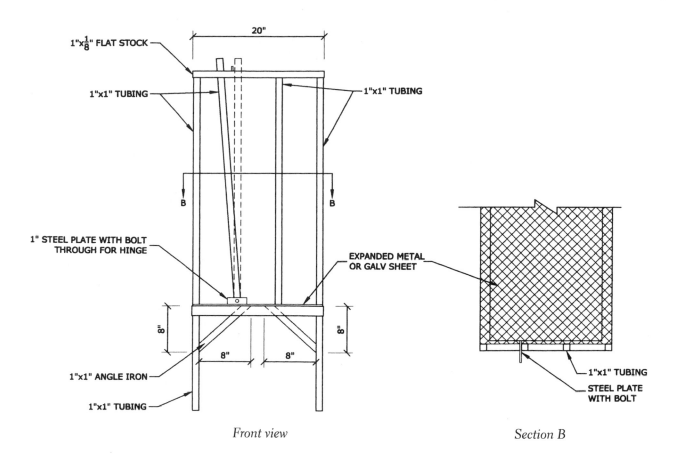

20"

1"x$\frac{1}{8}$" FLAT STOCK

1"x1" TUBING

1"x1" TUBING

B — B

1" STEEL PLATE WITH BOLT
THROUGH FOR HINGE

EXPANDED METAL
OR GALV SHEET

8" 8"

8" 8"

1"x1" ANGLE IRON

1"x1" TUBING

Front view

1"x1" TUBING

STEEL PLATE
WITH BOLT

Section B

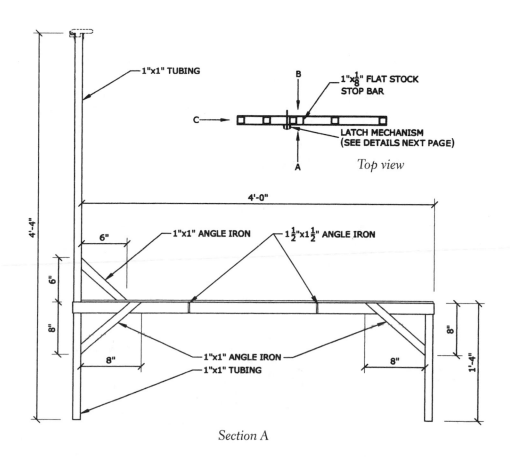

1"x1" TUBING

B

1"x$\frac{1}{8}$" FLAT STOCK
STOP BAR

C →

LATCH MECHANISM
(SEE DETAILS NEXT PAGE)

A

Top view

4'-0"

4'-4"

6"

6"

8"

8"

1"x1" ANGLE IRON

1$\frac{1}{2}$"x1$\frac{1}{2}$" ANGLE IRON

8"

1'-4"

8"

1"x1" ANGLE IRON

1"x1" TUBING

Section A

Latch mechanism detail

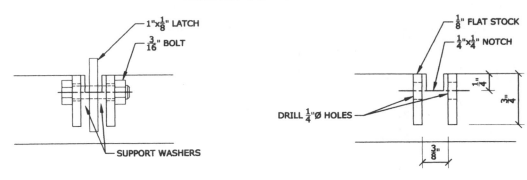

View A   View A without latch

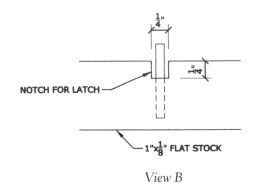

View B

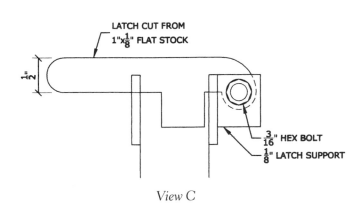

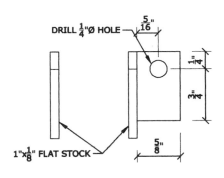

View C   View C without latch

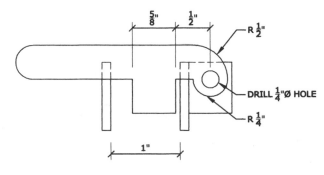

Cross section of View C at latch

Ø = diameter
GALV = galvanized

EQUIPMENT FOR PIGS

Feeders for pigs need to be sturdy. These designs work well for small herds, either in a building or on pasture.

Piglets need access to creep feeders for extra rations. Maintaining clean water is a real challenge with pigs because when it's hot out, they really like to roll in the mud — and they are smart enough to make their own mud holes at their watering point. For a small herd on pasture, a barrel waterer on skids works well. (*Never* use for this purpose barrels that contained toxic substances.)

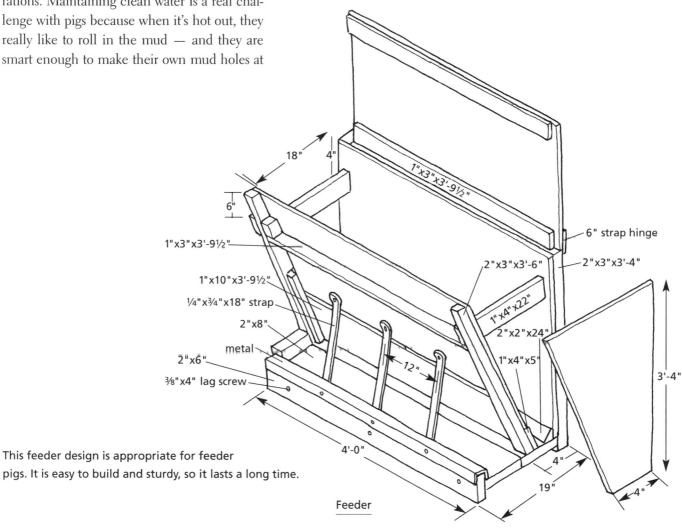

18" 4"

6"

1"x3"x3'-9½"

1"x10"x3'-9½"

¼"x¾"x18" strap

2"x8"

metal

2"x6"

⅜"x4" lag screw

1"x3"x3'-9½"

2"x3"x3'-6"

1"x4"x22"

2"x2"x24"

1"x4"x5"

6" strap hinge

2"x3"x3'-4"

3'-4"

12"

4'-0"

4"

19"

4"

Feeder

This feeder design is appropriate for feeder pigs. It is easy to build and sturdy, so it lasts a long time.

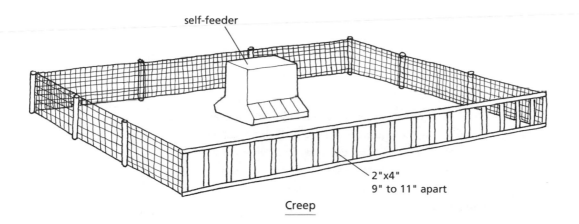

self-feeder

2"x4"
9" to 11" apart

Creep

This creep design works well for pigs on pasture. It allows small pigs access to extra feed without the pressure of sows and larger pigs pushing them away. Size the pen according to the number of smaller pigs you have.

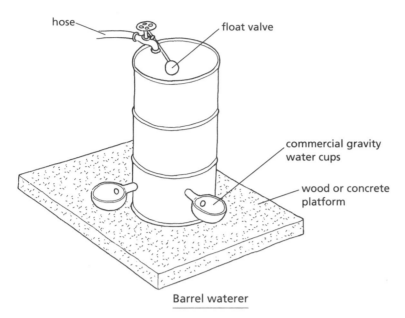

hose — float valve

commercial gravity water cups

wood or concrete platform

Barrel waterer

For a small pig operation, this barrel waterer works well. When full of water (the float valve keeps it full), pigs can't tip it over and they can drink whenever they want. Paint the barrel white and place it in the shade to keep the water cool.

There is nothing quite like the taste of pork from pigs raised on pasture. The meat is superior, with firm texture and less fat than pork that comes from industrial pig farms. When we farmed commercially, pasture pork was one of our top-selling products. From bacon and hams to steaks and chops, customers couldn't get enough.

If you will be farrowing on pasture in well-bedded portable huts during the warm months, you won't have to worry about cold piglets, and as a rule you won't have to worry about mother pigs rolling on their offspring. But if you will farrow indoors in winter, these become major concerns. Farrowing should take place in a box or farrowing crate that protects babies from being rolled on, and you will need a brooder.

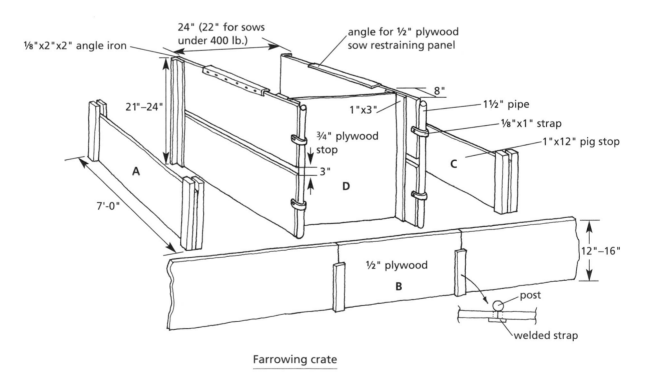

¹⁄₈"x2"x2" angle iron

24" (22" for sows under 400 lb.)

angle for ½" plywood sow restraining panel

21"–24"

8"

1½" pipe

1"x3"

¹⁄₈"x1" strap

1"x12" pig stop

¾" plywood stop

A

3"

D

C

7'-0"

12"–16"

½" plywood

B

post

welded strap

Farrowing crate

For farrowing indoors, use a farrowing crate like this one. The crate is constructed against a wall. The "pig stop" (pieces A, B, C) creates an area in which the piglets can move around and get away from the sow, while the crate itself (D) keeps the sow contained so she can't roll on them. The crate is elevated 8 to 10 inches above the floor, is secured to the wall, and rests on the pig stop in front.

Pig Brooder

This brooder is a good design for small pig operations. It provides a warm area in a pen where piglets are kept. They can move in and out of the brooder at will, according to the temperature. (See page 247 for plan credit.)

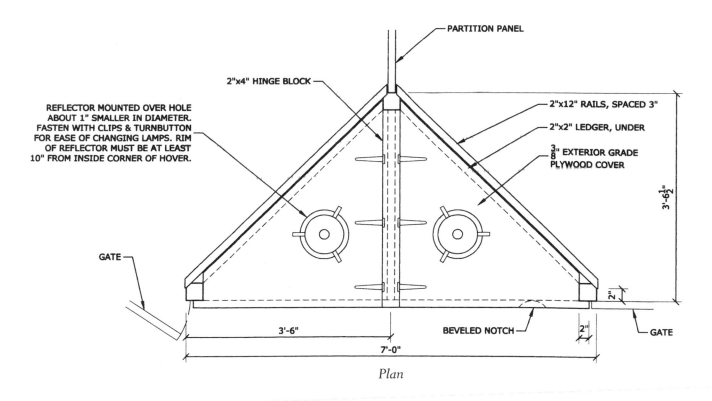

REFLECTOR MOUNTED OVER HOLE ABOUT 1" SMALLER IN DIAMETER. FASTEN WITH CLIPS & TURNBUTTON FOR EASE OF CHANGING LAMPS. RIM OF REFLECTOR MUST BE AT LEAST 10" FROM INSIDE CORNER OF HOVER.

PARTITION PANEL

2"x4" HINGE BLOCK

2"x12" RAILS, SPACED 3"

2"x2" LEDGER, UNDER

$\frac{3}{8}$" EXTERIOR GRADE PLYWOOD COVER

GATE

3'-6$\frac{1}{2}$"

2"

BEVELED NOTCH

2"

GATE

3'-6"

7'-0"

Plan

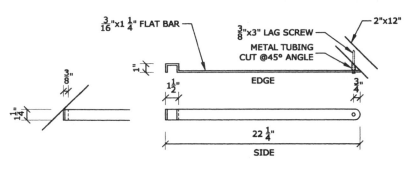

$\frac{3}{16}$"x1$\frac{1}{4}$" FLAT BAR

$\frac{3}{8}$"x3" LAG SCREW

2"x12"

METAL TUBING CUT @45° ANGLE

$\frac{3}{8}$"

1"

$\frac{3}{4}$"

EDGE

1$\frac{1}{2}$"

$\frac{3}{4}$"

1$\frac{1}{4}$"

22$\frac{1}{4}$"

SIDE

Locking bar details

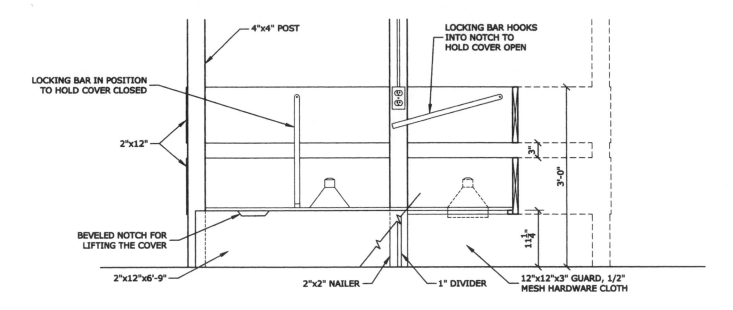

LOCKING BAR HOOKS
INTO NOTCH TO
HOLD COVER OPEN

4"x4" POST

LOCKING BAR IN POSITION
TO HOLD COVER CLOSED

2"x12"

BEVELED NOTCH FOR
LIFTING THE COVER

2"x12"x6'-9"

2"x2" NAILER

1" DIVIDER

12"x12"x3" GUARD, 1/2"
MESH HARDWARE CLOTH

3"

3'-0"

$11\frac{1}{4}$"

Elevation

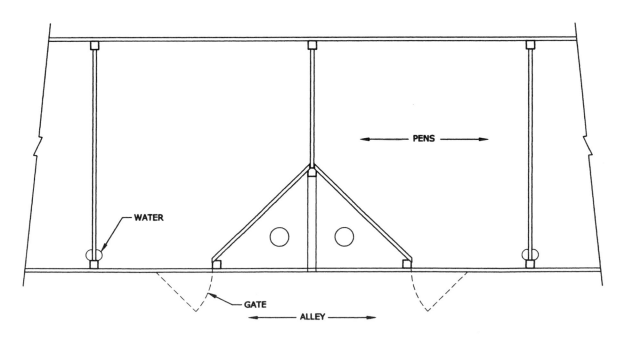

PENS

WATER

GATE

ALLEY

Pen arrangement

EQUIPMENT FOR CATTLE AND HORSES

Beef cattle and horses need primarily pasture and hay, supplemented with trace minerals and salt. They should have free access to these at all times. They can be supplemented with some grain fed in a bunk.

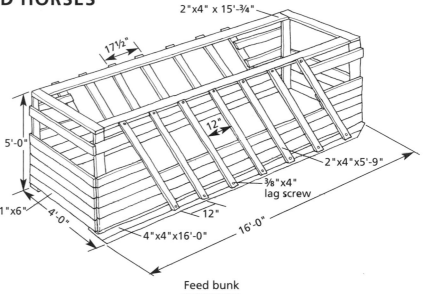

Feed bunk

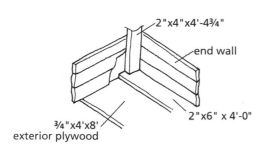

Bottom corner

Cows and most horses quickly learn to twist their heads to use this feed bunk, which is suitable for feeding hay or silage. You can use 1x6 lumber (as shown) or plywood for the walls. For cattle with large horns, like Scottish Highlands and Texas Longhorns, modify the design so there are three head slots per side and leave each slot open at the top.

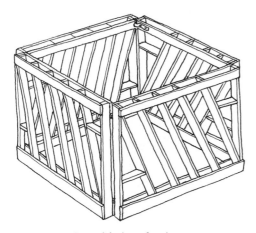

Portable box feeder

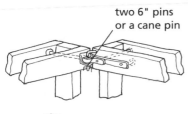

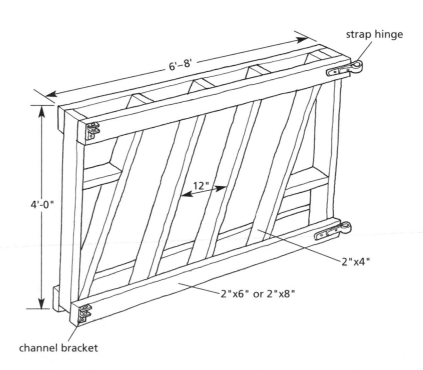

This portable box feeder can enclose a large round bale for feeding in the field. For large-horn cattle, modify the design so there are two head slots per side and leave each slot open at the top.

This feed bunk can be used for grain, minerals, and silage. The skids make it portable, allowing it to be pulled around the field.

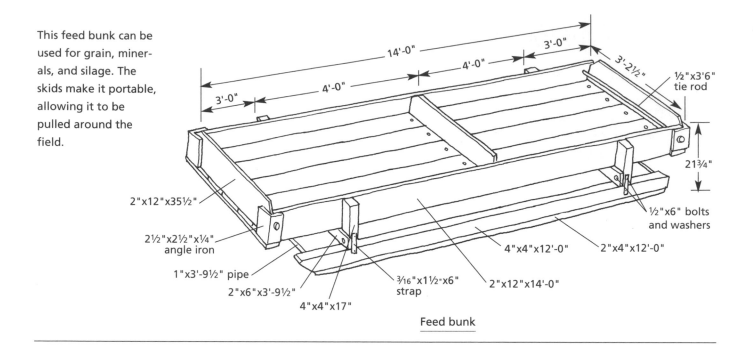

14'-0"

3'-0"

4'-0"

4'-0"

3'-0"

3'-2½"

½"x3'6" tie rod

21¾"

2"x12"x35½"

2½"x2½"x¼" angle iron

1"x3'-9½" pipe

2"x6"x3'-9½"

4"x4"x17"

³⁄₁₆"x1½"x6" strap

2"x12"x14'-0"

4"x4"x12'-0"

2"x4"x12'-0"

½"x6" bolts and washers

Feed bunk

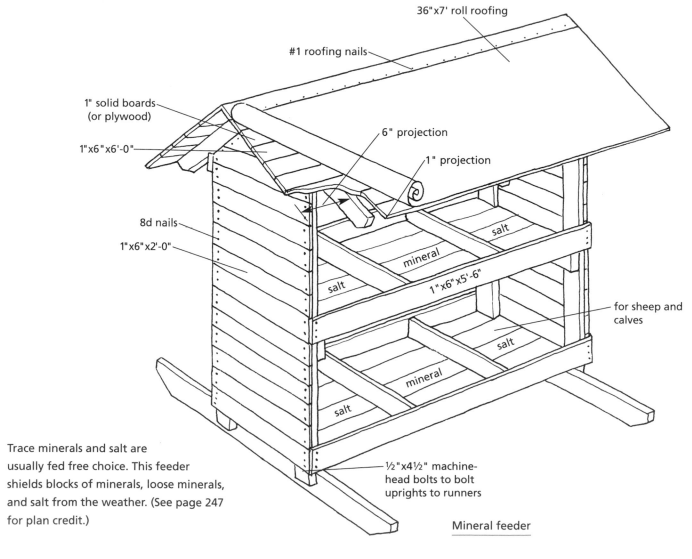

36"x7' roll roofing

#1 roofing nails

1" solid boards (or plywood)

1"x6"x6'-0"

6" projection

1" projection

8d nails

1"x6"x2'-0"

salt

mineral

salt

1"x6"x5'-6"

for sheep and calves

salt

mineral

salt

½"x4½" machine-head bolts to bolt uprights to runners

Trace minerals and salt are usually fed free choice. This feeder shields blocks of minerals, loose minerals, and salt from the weather. (See page 247 for plan credit.)

Mineral feeder

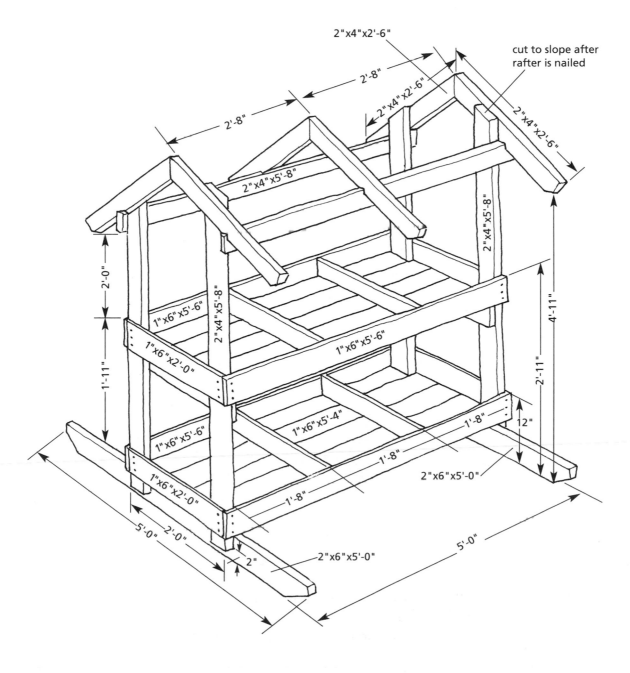

2"x4"x2'-6"

2'-8"

cut to slope after
rafter is nailed

2'-8"

2"x4"x2'-6"

2"x4"x2'-6"

2"x4"x5'-8"

2"x4"x5'-8"

2'-0"

1"x6"x5'-6"

2"x4"x5'-8"

4'-11"

1"x6"x2'-0"

1"x6"x5'-6"

1'-11"

2'-11"

1"x6"x5'-6"

1"x6"x5'-4"

1'-8"

12"

1"x6"x2'-0"

1'-8"

2"x6"x5'-0"

1'-8"

5'-0"

2'-0"

2"

2"x6"x5'-0"

5'-0"

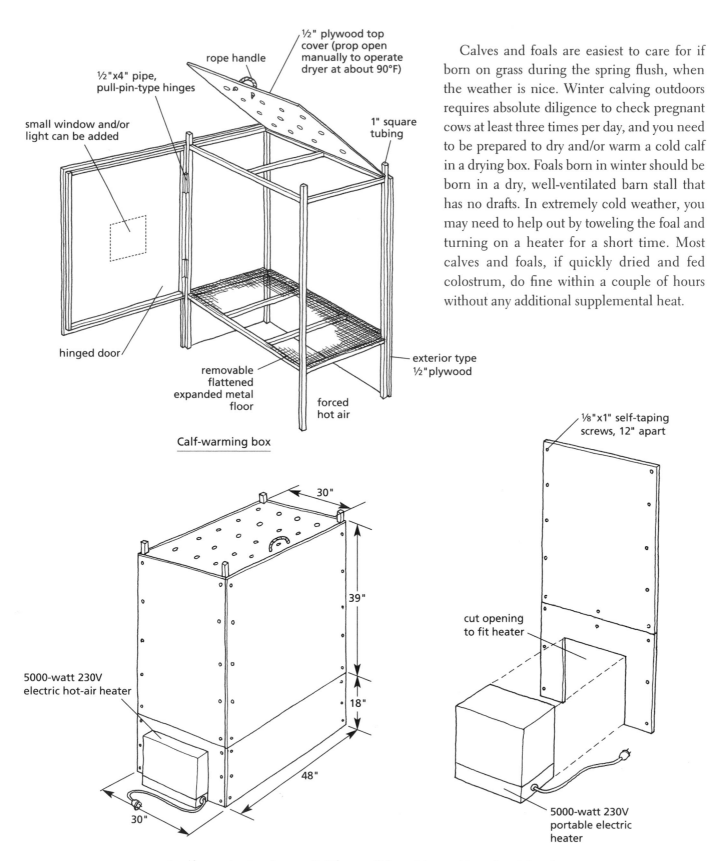

Calves and foals are easiest to care for if born on grass during the spring flush, when the weather is nice. Winter calving outdoors requires absolute diligence to check pregnant cows at least three times per day, and you need to be prepared to dry and/or warm a cold calf in a drying box. Foals born in winter should be born in a dry, well-ventilated barn stall that has no drafts. In extremely cold weather, you may need to help out by toweling the foal and turning on a heater for a short time. Most calves and foals, if quickly dried and fed colostrum, do fine within a couple of hours without any additional supplemental heat.

½" plywood top cover (prop open manually to operate dryer at about 90°F)

rope handle

½"x4" pipe, pull-pin-type hinges

1" square tubing

small window and/or light can be added

hinged door

removable flattened expanded metal floor

forced hot air

exterior type ½" plywood

Calf-warming box

30"

39"

18"

48"

30"

5000-watt 230V electric hot-air heater

⅛"x1" self-taping screws, 12" apart

cut opening to fit heater

5000-watt 230V portable electric heater

A calf-warming box is essential if you will be calving outdoors in winter. A cold calf can be placed in it for a couple of hours to dry off and get warm. Once the calf is dry and warm, it can go back outdoors with mom.

Horse owners (and cattle owners who show their animals) accumulate a fair amount of tack. These designs can assist with organizing your supplies.

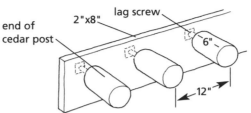

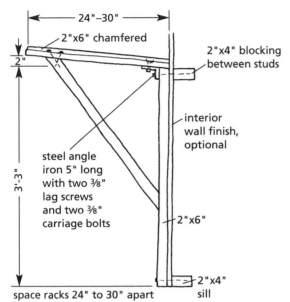

This simple rack stores halters and lead ropes, bellybands, and the other paraphernalia we horse enthusiasts collect. It doubles as a coat rack in our shed.

This is a handy rack design for storing saddle blankets. The cedar helps repel moths.

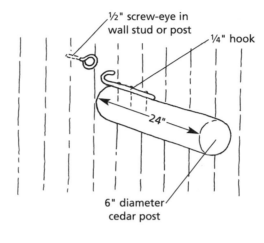

Racks

A

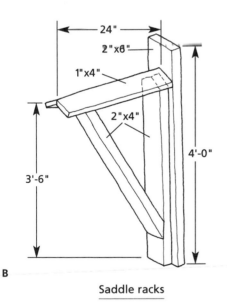

B

Saddle racks

Saddles are a big investment, and they last longer when stored on a saddle rack. These two designs (A, B) can help protect your investment.

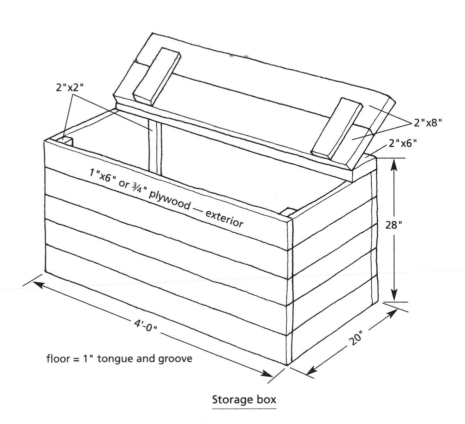

Storage box

This box will help you keep your tack room neat. Store odds and ends like brushes, farrier tools, and cleaning supplies in it. When the lid is closed, the box doubles as a bench.

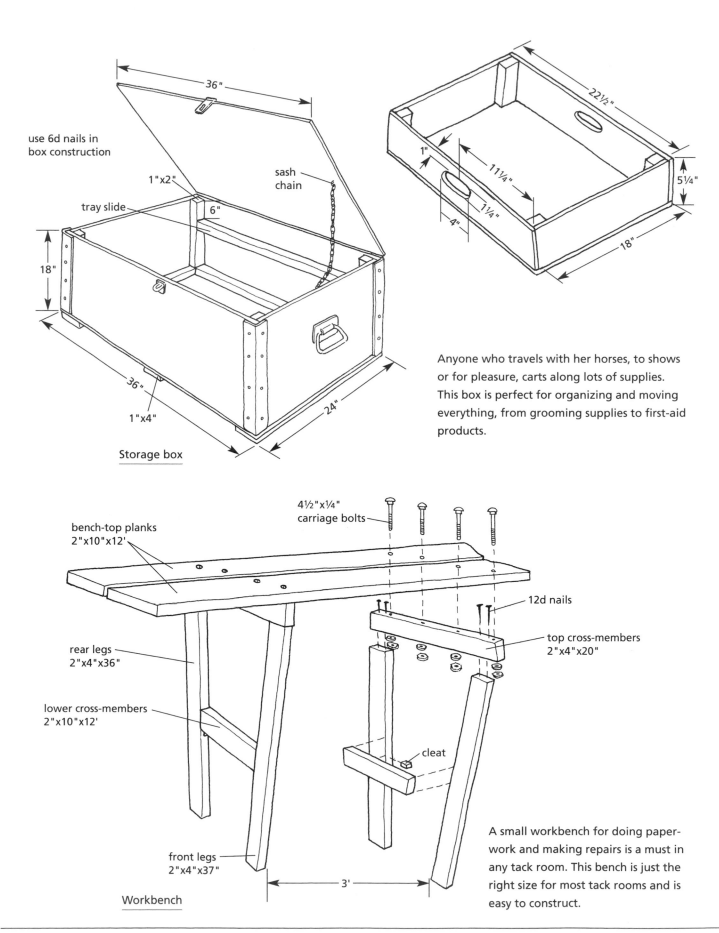

use 6d nails in
box construction

1"x2"

sash
chain

tray slide

6"

18"

36"

1"x4"

24"

Storage box

22½"

1"

11¼"

1¼"

4"

5¼"

18"

Anyone who travels with her horses, to shows
or for pleasure, carts along lots of supplies.
This box is perfect for organizing and moving
everything, from grooming supplies to first-aid
products.

4½"x¼"
carriage bolts

bench-top planks
2"x10"x12'

12d nails

top cross-members
2"x4"x20"

rear legs
2"x4"x36"

lower cross-members
2"x10"x12'

cleat

front legs
2"x4"x37"

3'

Workbench

A small workbench for doing paper-
work and making repairs is a must in
any tack room. This bench is just the
right size for most tack rooms and is
easy to construct.

Stall partitions for horses need to be sturdy and high, but they must allow adequate airflow. Tongue-and-groove boards topped with welded wire or commercially available stall guards (prefabricated metal rungs) make sturdy partitions and provide good ventilation.

STALL GUARDS

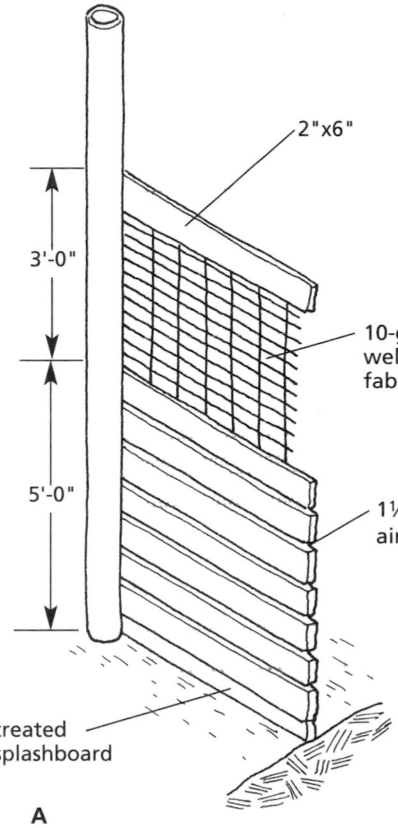

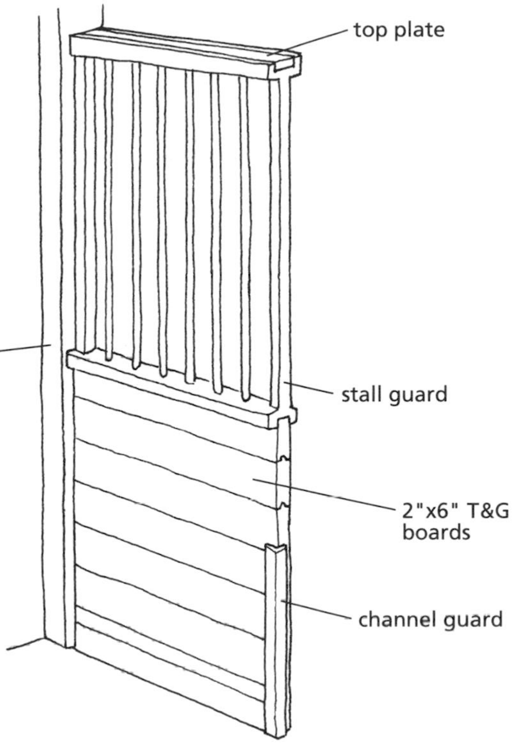

top plate

post

stall guard

2"x6" T&G boards

channel guard

B

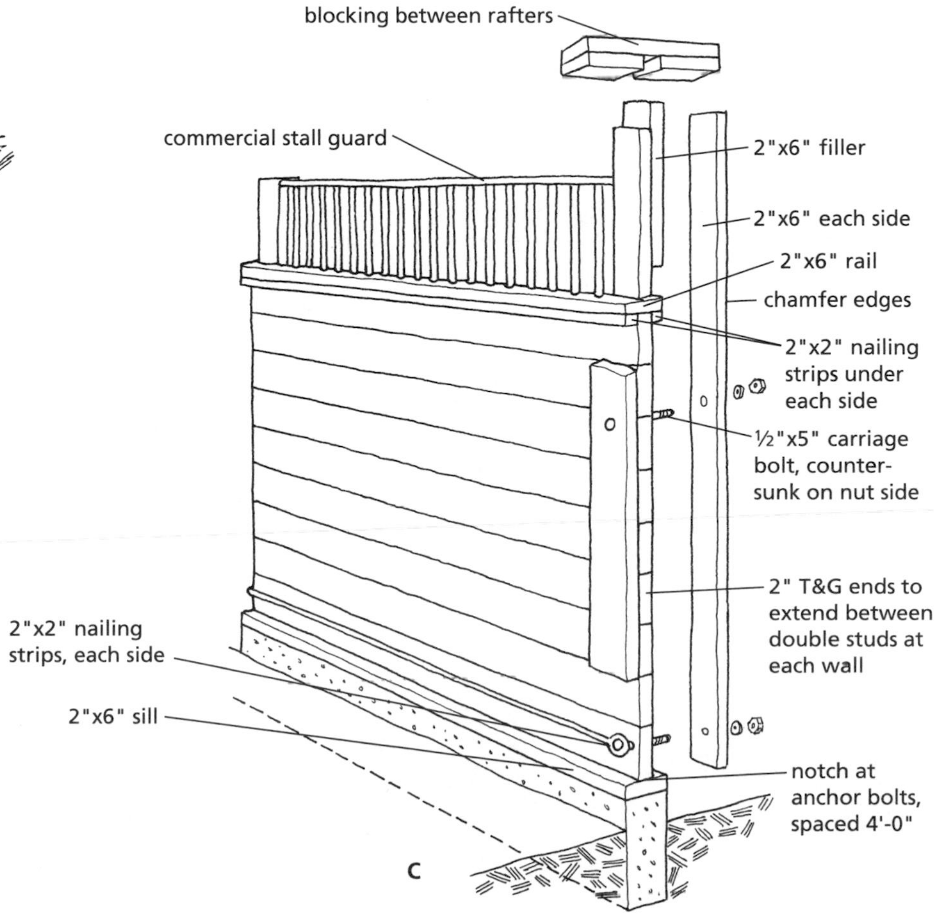

2"x6"

3'-0"

5'-0"

10-gauge welded-wire fabric

1½" space for airflow

treated splashboard

A

blocking between rafters

commercial stall guard

2"x6" filler

2"x6" each side

2"x6" rail

chamfer edges

2"x2" nailing strips under each side

½"x5" carriage bolt, counter-sunk on nut side

2" T&G ends to extend between double studs at each wall

2"x2" nailing strips, each side

2"x6" sill

notch at anchor bolts, spaced 4'-0"

C

Here are three possible stall guard designs. Tongue-and-groove boards (2x6 or 2x8) are the best choice for the lower half of stall partitions, and openings in the upper area are critical for adequate ventilation. (See page 247 for plan credits for illustrations A and C.)

III. CONSTRUCTION

For several years now, I have driven by a dream deferred — the forlorn, partially constructed frame of a very large barn. Unfortunately, such a sight is not all that unusual. Often, do-it-yourselfers and would-be builders don't understand just how significant an undertaking a major construction project can be: they run out of money or inspiration or both, often because they lack the skills necessary to complete the job.

If you plan to embark on a construction project but have limited experience, this section should help, but I urge you to start with a small project, like a shed, before trying your hand at a large barn or stable. It is much easier to learn techniques when building a simple 10'x10' shed than it is when trying to build a 3000-square-foot barn with a full wall foundation, complete upper story, and bathrooms and guest quarters.

Chapter 8, Before You Begin, provides general information on safety, tools, and materials. You may be tempted to skip over it, but don't: construction is dangerous work, and reading this chapter could save your life. It can also help you talk the talk when working with suppliers and subcontractors.

Chapter 9, Basic Construction, describes the nuts-and-bolts, how-to aspects of laying foundations, framing, and roofing. It also explains structural loads, variables that are critical to understand if you want the building to stand the test of time.

The last chapter, Final Steps, addresses finish work. It covers electrical systems, wiring, plumbing, and enclosing and finishing the shell of the structure. It also includes plans for building a cupola and advice on painting.

8. BEFORE YOU BEGIN

Construction can be a dangerous business, and often do-it-yourselfers are more vulnerable to injury than professionals because they don't have proper equipment or training. At the beginning of your project, be mindful of what you'll be doing and how.

At the start, you may be working in a ditch on water or sewer lines or on a foundation. Are you aware of the potential for cave-in and what precautions to take to protect yourself?

Before you know it, your ditch worries have disappeared, but next you'll be working with power tools. Their improper use can result in potentially fatal electrical shock or serious bodily harm. I don't know about you, but I'm kind of partial to my fingers.

Soon, you are climbing on ladders, scaffolding, or the structure itself, and it's a long way down to the ground. And metal ladders can contribute to electrical shock.

These are very real dangers that, according to the U.S. Occupational Health and Safety Administration (OSHA), make construction

the most dangerous occupation in terms of numbers of injuries and deaths. Construction accounts for 20 percent of all work-related deaths in the United States, even though it comprises only 11 percent of the workforce. Don't become a statistic: always think safety first.

HAZARDS
There are many hazards when working on construction sites. The most common, again according to OSHA, include the following:

- **Falls** from ladders, poorly designed scaffolding, and roofs, or tripping over tools and materials left around the work area.
- **Being struck by something;** for example, falling objects, excavating equipment, and vehicles.
- **Being caught in or between things.** This includes ditch cave-ins and being caught in moving parts of machinery or equipment that isn't properly protected by guards.

• **Electrical shock,** including electrocution. Electrical shock is often the result of using frayed or torn extension cords, or of improper grounding.

Ken and I know people who have been seriously injured or killed by each of these types of hazards. For example, Wayne, who is an electrician by trade, fell over a pile of boards someone left in the middle of a walkway while he was carrying boxes of supplies that obscured his vision; he broke both arms, missed a couple of months of work, and now, ten years later, suffers from arthritis in both arms as a result of the injury. Another one of our friends, Kurt, fell off a ladder and — at forty-seven years old — was killed when he hit his head on a rock below.

EXCAVATION SAFETY

More construction workers are killed in excavation cave-ins than in any other kind of accident. All excavations greater than 4 feet deep *absolutely* need to be protected from cave-in through the use of sloping, benching, or a shoring system. In all excavations greater than 3 feet deep, provide a ladder to enter and leave the cut; the ladder should extend at least 3 feet above the top of the cut.

Sloping the sides of an excavation is suitable for large areas, like a basement, but sloping should only be used in trenches less than 12 feet deep and in stable soils. The angle of the slope varies depending on the soil, though 2 of rise to 1 of run is the most common angle.

Benching is a process in which the sides of a ditch are stepped for greater safety; you can go to a depth of 20 feet with benching. Although sloping and benching are acceptable techniques for protection, you will be forced to excavate far more soil — and pay more — using these methods than you would using shoring.

Shoring uses wooden or metal panels to keep ditch walls from caving in; *trench boxes*

Ten Rules for Working Safely

1. Read the instructions before using a new tool or piece of equipment.
2. Don't wear jewelry or loose-fitting clothing while working.
3. Do wear appropriate personal protective gear, like steel-toed boots, hard hats (particularly if someone is working above you), goggles, ear protection, and cartridge-type respirators. Look for personal protective gear that has been approved by the National Institute for Occupational Safety and Health (NIOSH).
4. Make sure power tools and equipment have safety devices and guards in place. Don't carry power tools by cords, and keep cords clear of heat, sharp edges, and oil. Disconnect electricity before changing bits or blades.
5. Don't leave tools and materials lying around in the work area.
6. Use standard ladders and scaffolding that meet OSHA requirements. If you will be doing electrical work, use nonconductive ladders and scaffolding.
7. Make sure electrical circuits are not overloaded and that electrical systems are properly grounded.
8. Keep an adequate first-aid kit on-site (it should include emergency eyewash jars), and post numbers for emergency responders (fire, police, ambulance) near the phone. Consider taking a first-aid class before beginning a major project; classes are usually offered by local fire departments, community colleges, and hospitals.
9. Never work alone.
10. Use care when carrying sharp tools and materials. Don't carry sharp objects or nails in your pockets or mouth.

---

are commercially available shoring units that contractors slip into the excavation with a backhoe. As a rule of thumb, shoring begins to pay for itself in ditches of more than 7 feet deep. It's also the best option when you want to maintain a narrow cut because of adjacent buildings or utilities, narrow rights-of-way or easements, or because you just want to minimize site disturbance. Although it is advisable to hire a contractor who will use a trench box when shoring is required, you can create your own shoring using heavy plywood sheets or planks braced between ditch walls with 4x4s or jackscrews. When excavating deeper than 20 feet, have a registered, professional engineer design a system to stabilize the excavation site.

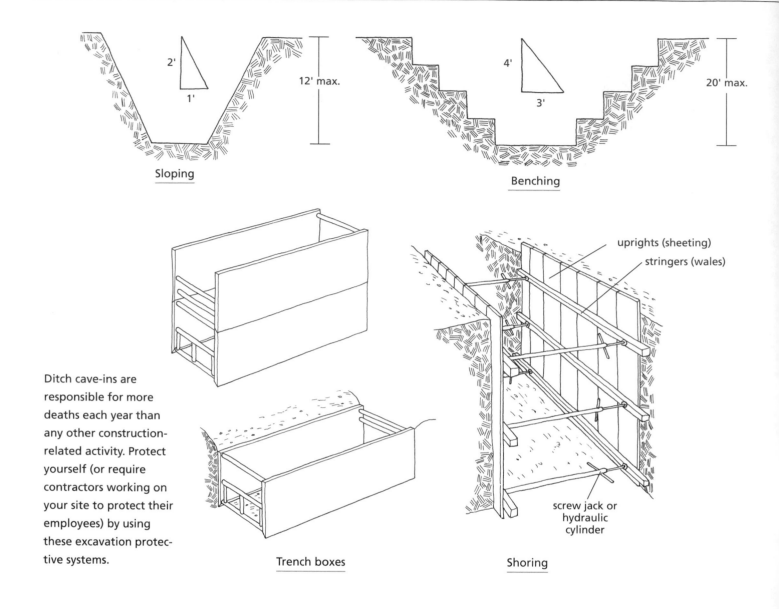

Sloping — 2' / 1' — 12' max.

Benching — 4' / 3' — 20' max.

uprights (sheeting)
stringers (wales)
screw jack or hydraulic cylinder

Trench boxes

Shoring

Ditch cave-ins are responsible for more deaths each year than any other construction-related activity. Protect yourself (or require contractors working on your site to protect their employees) by using these excavation protective systems.

USE THE RIGHT TOOL

The early Egyptians and Greeks knew a thing or two about geometry and trigonometry, which helped them construct amazing buildings that are still standing today. The key to their success was developing methods for creating structures that were square, straight, level, and true. Although materials have changed and we now have the benefit of power tools, there really aren't any significant differences in the building techniques developed more than twenty centuries ago and those we use today.

Having the right tool for the job and knowing how to use it will make your construction project go more smoothly and ultimately yield a better result. Trying to improvise with inappropriate or poor-quality tools will leave you frustrated.

A good variety of high-quality hand tools are required for any construction job and are relatively inexpensive. In fact, you could easily build small projects, like those in chapter 4 (Small & Portable Housing Projects), with no power tools. Although many an older building was built without the benefit of power tools, they are a real boon for larger projects, such as those in chapter 6 (Barns & Stables). Our own collection includes a wide variety of hand and power tools, some of which we could probably live without. Following is a list of absolute necessities.

• **Levels, squares, chalk line, and plumb bob.** These tools are essential if you want to build a structure that is level and plumb, straight, and true. They are critical for any construction project. At a minimum, you should have a 4-foot level and a 9-inch "torpedo" level (these can be either digital or the good old liquid-filled bubble variety); a large rafter square; a handheld combination square; a chalk line; and a plumb bob. You may also opt for some other levels, like a "carpenter's dumpy level" or a transit, though for occasional use these can be leased from a tool rental facility. There are other types of squares, such as speed squares and "try" squares, that supplement the combination square, but the best supplement is a bevel square, a good tool that allows you to duplicate funny angles without doing any math.

• **Hand tools.** Keep a well-stocked variety of hammers, handsaws, planes, chisels, screwdrivers, wrenches, pliers (including an electrician's pliers, which allows you to strip and cut wire), tape measures (a 25-foot tape is the most versatile, but a 100-foot tape is useful also), and other small hand tools. You will definitely need a pointed shovel and spade for digging, and a pry bar or rock pick may come in handy if you have rocky soil or will need to dig out tree roots. For handsaws, you'll want both a crosscut saw (to cut across the grain of the wood) and a ripsaw (to cut with the grain). A hacksaw is essential if you will be doing plumbing and electrical work (for cutting heavy wire, conduit). If you will be doing concrete work, you'll need a notched trowel, a wood float, and a groover.

• **Electric handheld circular saw and drill.** These are indispensable power tools. Table saws, radial arm saws, reciprocating saws, routers, grinders, planers, and other power tools are nice to have around if you have a generous tool budget and a good work space to use them in, but you can live without them; the circular saw and drill, however, will be in almost constant use. We use a 1.5 hp, 7¼-inch circular saw and a 3-amp, ½-inch drill. We also have a battery-operated drill that is about as handy as a tool could possibly be! Although not essential, reciprocating saws can replace a number of handsaws and are especially useful if you do remodeling.

• **Chain saw.** I don't think we have undertaken many projects in which the chain saw hasn't been put to good use. It makes cutting poles, beams, and other heavy wood products (like glue laminates) a cinch.

• **Ladders and scaffolding.** Most building projects are going to require you to get off the ground. Stepladders are fine if you can place them on firm, flat ground; otherwise, you're brewing trouble. Extension ladders come in lengths up to 50 feet long, but for most of us, a 16-foot or 20-foot extension ladder does the trick. Scaffolding is far safer to work from because it gives you a secure platform. Equipment-rental businesses offer scaffolding in several types. Putting together four of the stock panels animal owners use for corrals and pens, with heavy planks running between two, makes a good mid-height scaffold.

Shopping for Tools

Try on tools for size. If you typically wear gloves while working, take a pair with you when you plan to purchase tools and get a feel for the tool with and without gloves. Look for well-balanced tools with handles that are comfortable in your hand (not too thick, too thin, or too short), do not conduct electricity or heat, do not hurt your hand when you hold tight (you don't want sharp edges or finger grooves or ridges), and have a nonslip (rubber or plastic) handle. If you already own tools that are uncomfortable, consider getting a special plastic or rubber sleeve or a custom-grip kit for the handles.

When selecting power tools, choose a tool that is heavy enough to do the job but not so heavy it causes you to strain. You can tell how heavy-duty a power tool is by checking its amps or horsepower: the higher the number, the greater the power. Also, look for a tool that has a long "trigger" so you can use more than one finger to grip it.

Safety Always

To reduce the chances of injury, always wear appropriate personal protective equipment: wear hard hats when anyone is working above you, put on safety glasses when cutting or grinding, and protect your ears whenever using power tools.

When shopping for tools, purchase the best quality you can afford. Look for companies that have been in business for a long time and that offer good warranties. Craftsman tools from Sears are always high quality, and its hand tools, like hammers, saws, screwdrivers, and wrenches, are guaranteed forever. Craftsman, Porter Cable, and Skil all produce rugged but moderately priced power tools that last for years with the occasional use most homeowners put them to; Bosch, Dewalt, Makita, and Milwaukee make the "Cadillacs" of power tools, but you will pay more for such professional-level tools. If you are comparison shopping for tools and are wondering what professionals think of them, check out the online magazine *Tools of the Trade* (see resources for Web address); it offers reviews from contractors who work with the tools every day.

LUMBER PRIMER

Wood is probably the most common material used in construction. It's a gift from nature that is versatile and easy to work with, and it's a renewable resource. In fact, there are more trees today than there were twenty-five years ago — many of them "farmed," and many in forests that have grown overly thick due to current no-burn policies, especially in the West.

In spite of the good news that we have more trees, there is bad news in how some forests are managed. But the forest products industry is making strides in managing forests for both production and environmental interests with its *sustainable forestry initiative* (SFI), a pro-

WOOD CHARACTERISTICS

Species	Characteristics
Hardwoods	
Beech	A stiff hardwood with reddish color; used in flooring and trim
Birch	An even-grained wood that takes finish well and is used in doors and cabinets
Cherry	A very strong hardwood with reddish brown heartwood and dark brown rings; used in furniture and paneling
Hickory	Exceptionally tough and resilient, but shrinks significantly during drying; makes good ladder rungs, dowels, and poles
Locust, black	Very heavy and nearly indestructible; it resists decay, so it is often used for fence posts and rough construction
Maple	Strong and durable, with a variety of grain patterns; used in flooring, stairs, and paneling
Oak	Light colored, dense, and durable; used in flooring, stairs, and paneling
Softwoods	
Cedar	Decay resistant; used for paneling, shingles, and trim
Cypress	This swamp native is highly resistant to water rot and is used for siding, posts, trim, and exposed structural members
Douglas fir	Strong, dense softwood used in structural lumber
Ponderosa pine	Stable wood that takes stain well; used for trim, doors, windows, and paneling
Redwood	One of the most resistant to decay; used in siding and on decks and patios
Southern yellow pine	Stiff wood with a reddish tone (it gets its name from its pollen, which shows bright yellow in spring); used in structural lumber
Spruce	A softer group of softwoods; used in framing and for planks
White pine	Very soft woods of light color; used in cabinetry and for windows, doors, and paneling

gram many environmental organizations recognize as a step in the right direction.

One of the primary elements of the SFI is a third-party certification program that considers 118 core indicators before offering an SFI label. The indicators evaluate reforestation, water quality, and wildlife protection standards, among others.

Wood Talk

Wood comes in two general classifications — hardwoods, which are from slow-growing, deciduous trees (like beech, birch, cherry, oak, and maple), and softwoods, which come from faster-growing, evergreen species (such as cedar, fir, and pine). Hardwoods are more expensive and are typically used for finish work, furniture, and cabinetry; softwoods are more economical and make up the bulk of structural wood used in construction.

All mature trees have both heartwood and sapwood. The heartwood is the dense, dead wood at the center of a tree. It is the strongest part and is less vulnerable to decay than sapwood. *Sapwood* refers to the outer area of growth (young trees may have only sapwood). It carries sap (water mixed with minerals and nutrients), the fuel a tree needs to grow, between the soil and leaves. Sapwood is usually used for planks, studs, siding, and other components that will be protected from the elements or treated.

From Log to Lumber

Before you purchase lumber at a local yard or home store, it has gone through a number of steps. First, the log passes through a machine that strips away its bark. Then, depending on the size of the log, the sawyer establishes a cutting plan. Next the lumber is dried, typically in a kiln. Kiln drying drives off both free water (the water between wood fibers) and absorbed water (the water bound to cellulose in the wood-fiber cells) and yields a finished product with 5 percent to 19 percent retained moisture. As lumber dries, it shrinks, and as it shrinks, it may warp.

Finally, lumber is milled or planed. This last process yields the nice, smooth boards found at the lumberyard. It also reduces the size of the board from its "nominal size," or freshly sawn size, to its "finished" size. Thus, when you purchase a 2x4 at the lumberyard, those dimensions refer to its nominal size. The actual finished size of a 2x4 is only 1½x3½.

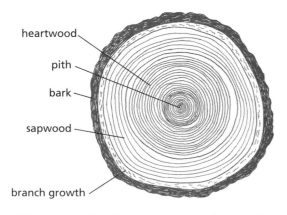

This cross section shows the growth characteristics of a tree. Heartwood is dense and does not tend to warp as much when it shrinks, so it is typically used for higher-quality boards and large structural beams. Sapwood is young wood, so it warps easily. It is cut for studs.

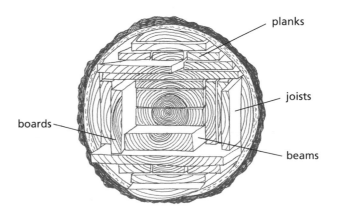

Every log is unique. Sawyers evaluate a log to determine the best cutting plan for yielding usable lumber.

Native, or Green, Lumber

Often available from sawmills in forested regions, native, or green, lumber is rough sawn and not kiln dried. The advantage of purchasing native lumber is it is less expensive. The disadvantages relate primarily to moisture content: it is usually heavy to work with, and as it dries it's subject to greater splitting and cracking than kiln dried lumber. It also has slightly uneven sizing, and the crown, or the rounded side of the board, can be difficult to see. (Boards are usually laid with the crown up for nailing.)

Depending on the region where it was cut and its size, the moisture content (MC) of a freshly felled tree typically runs in the range of 25 to 55 percent, though it can reach 250 percent. (To determine the MC, the weight of water in the wood is divided by the weight of dry material. For example, if a green-cut board

AMERICAN STANDARD LUMBER SIZES FOR YARD AND STRUCTURAL LUMBER CONSTRUCTION

Item	Nominal inches (Thickness)	Dry inches	Dry mm	Green inches	Green mm	Nominal inches (Face Width)	Dry inches	Dry mm	Green inches	Green mm
Boards	1	¾	19	25/32	20	2	1½	38	19/16	40
	1¼	1	25	1 1/32	26	3	2½	64	2 9/16	35
	1½	1¼	32	1 9/32	33	4	89	3½	90	3 9/16
						5	4½	114	4⅝	117
						6	5½	140	5⅝	143
						7	6½	165	6⅝	168
						8	7¼	184	7½	190
						9	8¼	210	8½	216
						10	9¼	235	9½	241
						11	10¼	260	10½	267
						12	11¼	286	11½	292
						14	13¼	337	13½	343
						16	15¼	387	15½	394
Dimension	2	1½	38	1 9/16	40	2	1½	38	1 9/16	40
	2½	2	51	2 1/16	52	3	2½	64	2 9/16	65
	3	2½	64	2 9/16	65	4	3½	89	3 9/16	90
	3½	3	76	3 1/16	78	5	4½	114	4½	117
	4	3½	89	3 9/16	90	6	5½	140	5½	143
	4½	4	102	4 1/16	103	8	7¼	184	7½	190
						10	9¼	235	9½	241
						12	11¼	286	11½	292
						14	13¼	337	13½	343
						16	15¼	387	15½	394
Timbers	≥5	½ in. off	13 mm off	½ in. off	13 mm off	≥5 off	½ in. off	13 mm off.	½ in off	13 mm off

From Forest Products Laboratory, *Wood Handbook: Wood as an Engineering Material*, Gen. Tech. Rep. FPL-GTR-113 (Madison, WI: U.S. Department of Agriculture, Forest Service, Forest Products Laboratory), 5–11.

weighs 30 pounds before drying and 20 pounds after, the MC equals 10 pounds water divided by 20 pounds of dry material, or 50 percent.)

In spite of its drawbacks, native lumber is great for small projects like chicken coops and sheds and works very well as siding for larger projects; it can be used in board-and-batten siding with no drying. If you have access to a woodlot or local forests, you may be able to find a traveling operator who will come right to your place and cut lumber to your specifications. When selecting trees for cutting on your own land, it is a good idea to contact your state's forestry department. For little or no charge, a trained forester will come and help you select the best trees for your purpose.

If you purchase from a local mill (few local mills dry their lumber) or use wood from your own woodlot, you can air-dry the lumber. Because air-drying can take anywhere from a couple of months to close to a year, depending on your climate, it requires some advance planning.

NOMINAL AND ACTUAL LUMBER SIZES

Nominal size (in.)	Actual size (in.)	Nominal size (in.)	Actual size (in.)
1x2	¾x1½	2x2	1½x1½
1x4	¾x3½	2x4	1½x3½
1x6	¾x5½	2x6	1½x5½
1x8	¾x7¼	2x8	1½x7¼
1x10	¾x9¼	2x10	1½x9¼
1x12	¾x11¼	2x12	1½x11¼
⁵⁄₄x2	1x1½	4x4	3½x3½
⁵⁄₄x4	1x3½	6x6	5½x5½

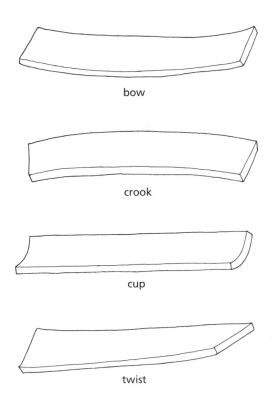

bow

crook

cup

twist

As lumber dries, it shrinks, and as it shrinks it can warp, crack, or split. These illustrations identify some of the typical problems encountered with lumber.

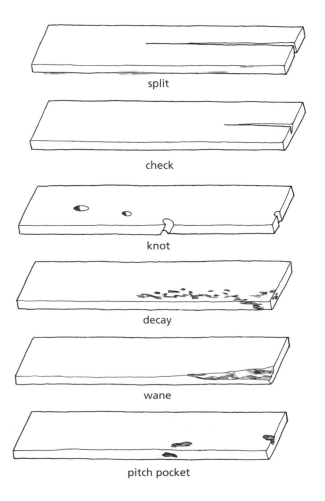

split

check

knot

decay

wane

pitch pocket

Drying Native Lumber

Air-drying lumber can take anywhere from thirty days to more than a year, depending on species, board thickness, air circulation, temperature, and relative humidity. To reduce checking (a lengthwise gap or crack) and splitting, begin drying in the fall and early winter, when temperatures are cool, and plan to use the lumber late the following summer.

The single most important consideration in successful air-drying of lumber is proper stacking. Stacks should be set on well-drained sites in an area free of brush and trees or sawdust and weeds.

Build a foundation by placing concrete blocks under the ends of railroad ties or beams, with rows 3 feet apart. Put one layer of lumber down, leaving a 1-inch gap between boards for circulation. Place "stickers," which are 1x2 boards, across the pile, then add another layer. As with the foundation, the stickers should be 3 feet apart.

Keep layering boards and stickers until you have a stack 3 feet to 4 feet tall. Add a layer of 4x4s on top of the stickers to support a sheet of wired-down galvanized metal used as a roof.

Days Required to Air-Dry Lumber[a]

Species	Days
American beech	70–200
Ash, white	60–200
Aspen	50–150
Basswood	40–150
Birch, yellow, sweet	70–200
Butternut	60–200
Cedar, eastern red	60–90+[b]
Cedar, northern white	80–130[b]
Cherry, black	70–200
Cottonwood, eastern	50–150
Fir, balsam	150–200[b]
Fir, Douglas	20–180
Hemlock, eastern	90–200
Hickory	60–200
Locust, black	120–180
Maple, hard (sugar, black)	50–200
Maple, soft (red, silver)	30–120
Oak, red	70–200
Oak, white	80–250
Pine, eastern white	60–200
Pine, red (Norway)	40–200
Pine, southern yellow	30–150
Poplar, yellow	40–150
Spruce, red and white	30–120
Sycamore	30–150
Walnut	70–200

[a]Twenty percent moisture content
[b]Species usually kiln dried
From Forest Products Laboratory, *Air-Drying of Lumber,* Gen. Tech. Rep. FPL-GTR-117 (Madison, WI: U.S. Department of Agriculture, Forest Service, Forest Products Laboratory), 24.

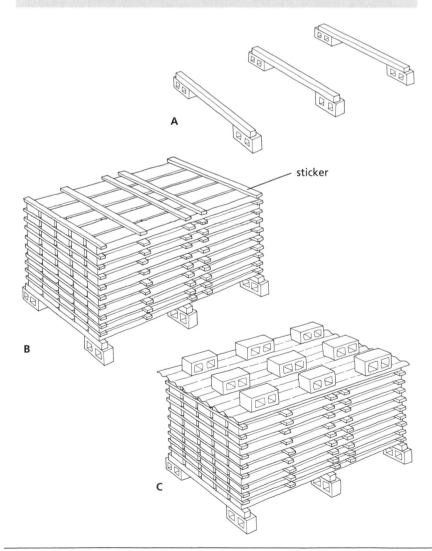

A

sticker

B

C

To dry lumber at home, create a base (A), stack with "stickers," separating the boards (B), and finish with a roof of galvanized metal to protect them from rain and snow (C).

Grades of Lumber

Any lumber you purchase from a commercial supplier has been graded. The grading process is directed by several different independent agencies around the United States based on standards adopted by the National Hardwood Lumber Association or the Department of Commerce. These graders take into account characteristics like moisture content, strength, resistance to stress, general appearance, and flaws that could affect strength (knots, cracks, warps).

Building codes require the use of graded lumber for structural components (though some areas exempt agricultural buildings from this requirement). The grades for structural lumber are:

- **select structural,** which has superior strength, stiffness, and resistance to stress, complemented by good appearance that makes it suitable for exposed beams, stringers (horizontal members that connect other members; also supports for treads in stairs), and decking; Douglas fir and hemlock are used for most structural lumber;
- **No. 1, or construction grade,** which is close to select but may have some small knots or considerations that affect appearance such as poor color or staining. No. 1 can be used in all the same applications as select, but it is usually used where it won't be exposed;
- **No. 2, or standard grade,** which has more knots but they are tight. This is typically used for studs, floor framing, and roof framing (joists, rafters, ridgeboards);
- **No. 3, or utility grade,** which has more defects but is suitable for sills and plates and for sheathing on roofs; and
- **economy,** which is only for nonstructural uses like temporary bracing and stakes.

For animal housing projects, standard grade, or No. 2, will be suitable for almost everything you do, and the cost difference between it and utility grade is marginal.

Treated Lumber

Pressure-treated lumber is a specialty product intended for use where the wood is likely to absorb moisture. It is soaked under pressure with insecticide and fungicide sprays. Pressure-treated lumber is usually used on decks and piers and other points where it might be affected by dampness.

One problem with using pressure-treated lumber, particularly in animal housing, is the potential toxicity from the chemicals used to treat the lumber. Avoid using it in locations where animals could ingest it. Horses are especially notorious for "cribbing," or chewing on exposed wood out of boredom, but other animals also lick or chew on exposed wood.

Wood Products

Most projects involve the use of engineered products made from wood, such as plywood, glulams (beams made by glue-laminating boards together), wood I-joists, oriented strandboard (or OSB, which is made from layers of long, narrow parallel strands, the orientation of the strands being alternated from one layer to the next), and structural particleboard.

CCA

In recent years, chromated copper arsenate (CCA) was the most commonly used chemical wood preservative, but as of January 1, 2004, the U.S. Environmental Protection Agency prohibited its use in wood intended for most noncommercial applications. New products are being marketed as safe alternatives, but I recommend avoiding treated lumber in any form, because animals may ingest the chemicals used in pressure treating. We have used pressure-treated wood only in the buried piers of our animal housing projects.

If you work with pressure-treated lumber, minimize your exposure by wearing gloves, along-sleeved shirt, long pants, and eye protection. When cutting it, work outdoors if possible and wear a mask.

TIP

National home stores such as Home Depot and Lowes offer a 10 percent discount off your first purchase if you use one of their credit cards. Ask for a bid for the entire building package, and open a new account to get additional savings.

These products are designed for specialized applications and to be as strong as or stronger than ordinary lumber.

Plywood is probably the most common and versatile of these products. It's made by gluing together thin layers of wood, called *plies*. To enhance its strength, it is always constructed with an odd number of layers, and the direction of the wood grain is alternated from one layer to the next. Layers may consist of a single ply or several plies glued together with the grain going in the same direction. The more plies there are, the stronger the sheet of plywood. As with lumber, there is a hardwood type, typically used in cabinetmaking and other specialty projects, and a softwood type that's considered the general construction type. Construction-type plywood comes in 4-foot by 8-foot and 4-foot by 10-foot sheets, and is available in:

- **interior,** which does not use waterproof adhesives, and which can be made with lower-quality, class D veneers, or plies;

- **exposure 1,** which is specifically for use in roof sheathing and interior walls; it uses waterproof adhesives but may use class D veneers;
- **exposure 2,** which is used in subfloors; it uses a class C or better veneer and adhesives that are moderately resistant to water; and
- **exterior grades,** which require waterproof adhesives and class C or better veneers for exposed wall surfaces.

Plywood is graded and bears grade stamps giving detailed information about the product, such as its thickness, the type of glue used to create it, its veneer grades, and whether it is suitable for interior or exterior use.

Composites such as strandboard and particleboard are made from fibers of wood instead of thin slices of wood. Some have plastics or other materials — like agricultural waste — added. The advantages of these products are that they are significantly less expensive than lumber or plywood and they are light and easy to work with. The disadvantages include their being less durable structurally and more prone to water damage, and subject to off-gassing, a problem associated with all glued wood products.

Off-gassing occurs when the chemicals — particularly formaldehyde — used in the production of glued wood products volatilize. Some individuals may have allergic reactions while working with these products or even when spending time in a completed structure where some gas remains. Symptoms associated with off-gassing can include nausea, vomiting, abdominal pain, and diarrhea. When the reaction is allergic, symptoms may include headaches, fatigue, minor respiratory irritation, and watery eyes. Infants, the elderly, and individuals with specific allergic reactions are most vulnerable. To minimize problems, ensure adequate ventilation while working with these products and for at least sixty days after installation.

Lumber Terms

Here are some of the terms you will encounter when talking lumber:

Board foot. A unit measure of lumber or sawlogs represented by a board 12 inches long by 12 inches wide by 1 inch thick, or the cubic equivalent. Some examples:

- A 1-foot-long piece of 1x12 lumber = 144 cubic inches (12 x 1 x 12)
- A 2-foot-long piece of 1x6 lumber = 144 cubic inches (24 x 1 x 6)
- A 3-foot-long piece of 1x4 lumber = 144 cubic inches (36 x 1 x 4)
- A 1-foot-long piece of 2x6 lumber = 144 cubic inches (12 x 2 x 6)
- A 1.5-foot-long piece of 2x4 lumber = 144 cubic inches (18 x 2 x 4)

Check. A lengthwise gap or crack.

Crosscut. To cut a piece of wood at a right angle to the grain.

Nominal. The size of freshly sawn lumber before drying and planing.

Pressure-treated. Treated with chemicals under pressure to reduce rot caused by water, insects, or fungus.

Square. An area of wood (or other materials, like roofing) that covers 100 square feet.

Warp. Distortion in board shape caused by drying.

HOLDING IT TOGETHER

Head into any well-stocked lumberyard, home store, or hardware store and you'll quickly be overwhelmed by the variety of fasteners available to hold your structure together. Nails are probably the most common, but there are also spikes, staples, screws, bolts, lags, truss plates, framing connectors, and adhesives. Each has its best application.

Nails are still designated by the "penny," a traditional term referring to the cost of a hundred nails. The letter *d* following a number tells you the penny size (2d equals 2 penny nails). Nails 4 inches or longer are typically called *spikes*.

The most widespread type of nails is "bright," or common, nails. These are just plain, smooth nails, but there are also a number of specialty types, such as cement coated (which are actually coated with a resin compound), zinc coated, annularly threaded, helically threaded, or barbed. The cement nail is designed to be easily driven but difficult to withdraw. Zinc-coated nails are used where corrosion could be an issue. Threaded and barbed nails are preferred for use when wood has been pressure-treated with preservatives, because they have much higher withdrawal resistance; the preservatives tend to act as a lubricant that allows common nails to work themselves out easily.

Common nails are good for most basic construction. Use a nail size slightly smaller than the wood you are nailing together: the nail should not come all the way through the second board but should penetrate it to at least half its width.

There are nails for special purposes — for example, roofing nails, flooring nails, ring-shank nails (used for subflooring), and finish nails. Duplex, or double-headed, nails are used for temporary nailing (the second head makes them easy to remove); they are used extensively in concrete forming or for nailing temporary bracing in place during framing.

Although staples are widely used in prefabricated construction (of modular homes, for example), they have limited use in most on-site construction projects. The applications for which they are well suited, however, include insulation, vapor barriers, roofing paper, and

NAIL SIZES				
Size	Length (in.)*	Gauge no.	Head diameter	Approx. no./ lb.
2d	1	15	11/64	847
3d	1¼	14	13/64	543
4d	1½	12½	¼	294
5d	1¾	12½	¼	254
6d	2	11½	17/64	167
7d	2¼	11½	17/64	150
8d	2½	10¼	9/32	101
9d	2¾	10¼	9/32	92
10d	3	9	5/16	66
12d	3¼	9	5/16	61
16d	3½	8	11/32	47
20d	4	6	13/32	29
30d	4½	5	7/16	22
40d	5	4	15/32	17
50d	5½	3	½	13
60d	6	2	17/32	10

*Length from underside of head to tip of point.

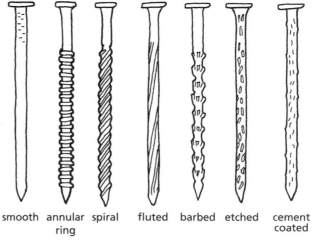

smooth   annular ring   spiral   fluted   barbed   etched   cement coated

There are dozens of types of nails, each with a specific purpose. Smooth nails are the most common; others are designed to make removal difficult or for use in highly corrosive environments.

Below are some of the most common screws, bolts, and nails you will find at your local hardware or building supply store. Using the correct one for the job will make your work easier and your building stronger.

other similarly light materials that need to be held in place temporarily until another layer goes on. Structurally rated staples can be used to apply sheeting products like OSB, but these require a staple gun.

Screws and bolts can be substituted for nails in many applications, but they are more expensive and slower to work with. The one application where they have a real advantage is on decks, where they are much less likely than nails to work themselves out. With the exception of anchor bolts or J-bolts, which are used to attach the sill plate (see chapter 9) to the foundation, bolts are rarely used in rough

carpentry, though if you are connecting large structural timbers to each other, running carriage bolts through drilled holes may be easier than trying to nail spikes.

Framing connectors are required by code for certain applications, such as connecting joists to girders and headers, but even if they aren't required by code, they are convenient for connecting joists, rafters, and trusses. Types of connectors include anchors, which connect members at right angles to each other; splice plates, which connect two members in line with each other; and split-ring connectors, which fasten members in a roof truss.

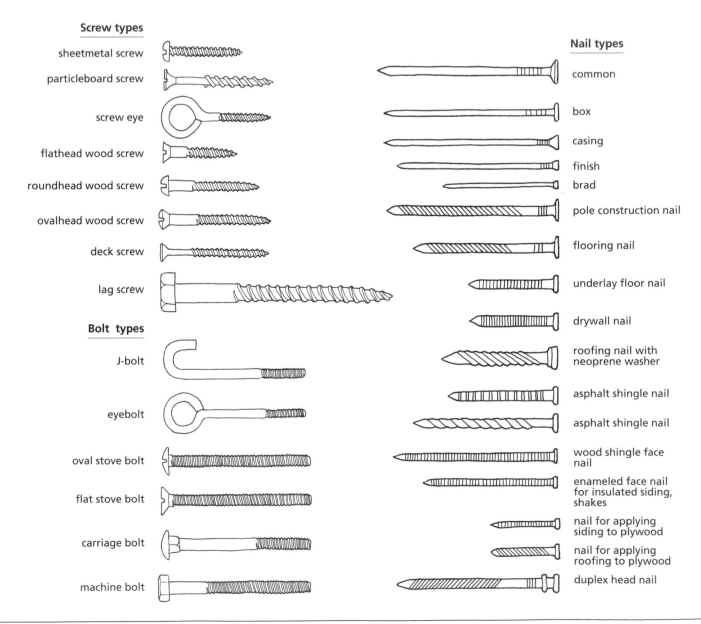

Screw types

sheetmetal screw

particleboard screw

screw eye

flathead wood screw

roundhead wood screw

ovalhead wood screw

deck screw

lag screw

Bolt types

J-bolt

eyebolt

oval stove bolt

flat stove bolt

carriage bolt

machine bolt

Nail types

common

box

casing

finish

brad

pole construction nail

flooring nail

underlay floor nail

drywall nail

roofing nail with neoprene washer

asphalt shingle nail

asphalt shingle nail

wood shingle face nail

enameled face nail for insulated siding, shakes

nail for applying siding to plywood

nail for applying roofing to plywood

duplex head nail

FRAMING CONNECTORS

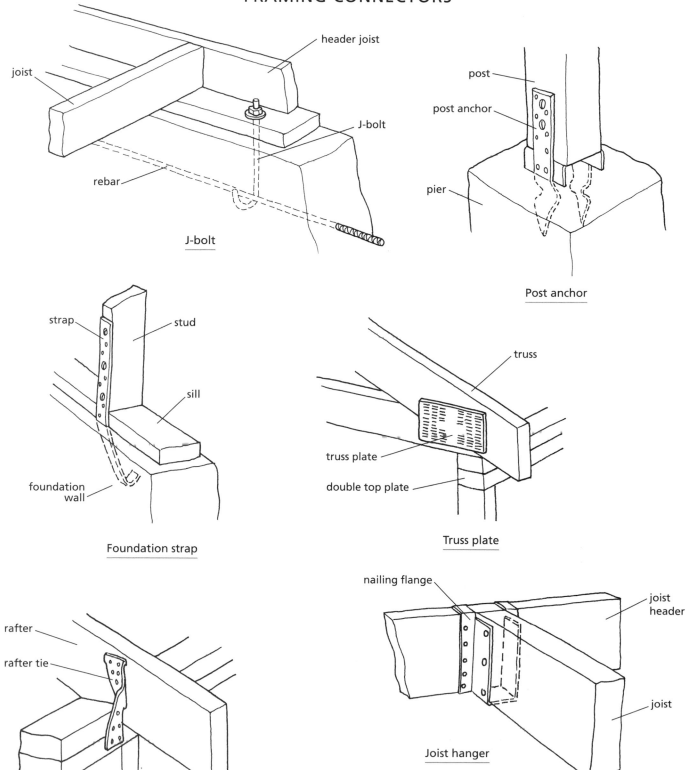

joist

header joist

J-bolt

rebar

J-bolt

post

post anchor

pier

Post anchor

strap

stud

sill

foundation wall

Foundation strap

truss

truss plate

double top plate

Truss plate

rafter

rafter tie

stud

Rafter tie

nailing flange

joist header

joist

Joist hanger

These framing connectors are the most common. Although you may not be required by code to use connectors, they are convenient and easy to use, making them a wise investment.

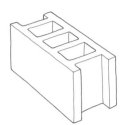

Stretcher (3 core)

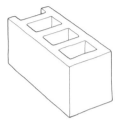

Corner

Half stretcher

Half corner

4" or 6" partition

Stretchers are the most common concrete blocks and come in full and half sizes; corner blocks are flat on one edge. Interior partitions can be built with full-width blocks, but code also allows the use of narrow blocks.

CONCRETE AND CEMENT

Concrete is usually used for foundations and floors but can also be used for walls. It's made basic components: cement, water, and washed aggregate (sand and/or stone). It has tremendous compressive strength, meaning it can support lots of weight, but low tensile strength, meaning it tends to bend and twist. To reduce bending and twisting, it is reinforced with wire mesh or reinforcing bar (rebar).

Portland cement is made from powdered and burned limestone or clay, and gypsum; these ingredients are mixed with water to create a paste that hardens as it dries and can glue aggregate (pieces of rock) together. The type and proportion of aggregate varies depending on locale and specified use. The quality of concrete is dependent on the ratio of cement to water to aggregate (a thicker paste produces stronger concrete than a watery paste). Mortar is cement with additional lime and screened sand, and it is used for laying up a concrete block (or brick) wall.

For very small jobs, you can purchase premixed bags of dry concrete. They come in half-cubic-foot sizes, so two bags would fill a slab approximately 4 inches thick by 20 inches square. For any sizable job, it's best to order concrete from a ready-mix operator. Most operators can help you determine the best mixture and the amount you will need for the job. They take orders in cubic yards, and 1 cubic yard is equal to 27 cubic feet.

Concrete is poured into forms, which can be made of wood. The ground beneath the pour typically needs to be cleared of weeds, trash, or sod, and it should be well compacted. If necessary for drainage, a layer of sand or gravel can be applied to the ground and should be leveled and tamped well before concrete is poured. As the concrete dries, joints are added to prevent cracking. Bigger pours require rebar or welded-wire reinforcing panels for strength and load distribution.

Concrete needs time to cure after it has been poured; the time required varies depending on weather and the thickness of the pour. It cures more quickly in hot weather (but it can cure too quickly), and more slowly when it's cold out. The ideal temperature range for concrete is between 50°F and 70°F. Avoid pouring when temperatures are consistently below 40°F or consistently above 85°F, though if you must work outside these ranges, there are techniques and additives that a ready-mix operator can help you with to improve the final pour. For example, freezing during the cure period results in weakened and damaged concrete, but you can add calcium chloride, which speeds curing time, and insulate the pour with commercially available insulating blankets or with tarps and hay. During extreme heat periods, keep the concrete damp for at least five days after the pour so it doesn't dry too quickly. Covering the pour with plastic sheeting will also help keep it damp. Commercially available curing compounds can be added to the concrete or painted on its surface to help it cure in extreme temperatures.

In areas where freezing temperatures are normal, concrete that will be exposed above ground will benefit from a process called *air entrainment*. The process produces millions of tiny air bubbles in the concrete through a chemical reaction. Air entrainment reduces

Caution with Wet Concrete

Because of its extremely alkaline nature, wet concrete can cause chemical burns to the skin. When working with it, wear protective eyewear, rubber boots, and rubber gloves at all times. Never walk or stand in concrete while spreading it unless you are wearing boots that come up at least several inches above the level of the wet concrete. If you splatter it on yourself, rinse it off immediately.

JOINTS, WIRE MESH, AND REBAR

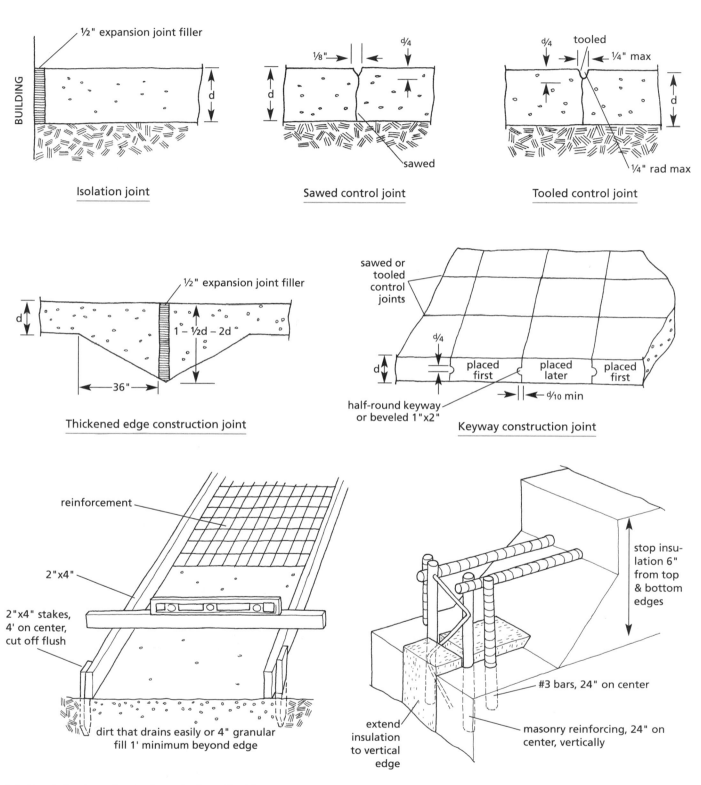

½" expansion joint filler

BUILDING

d

Isolation joint

⅛"   d/4

d

sawed

Sawed control joint

d/4   tooled   ¼" max

d

¼" rad max

Tooled control joint

½" expansion joint filler

d

1 – ½d – 2d

36"

Thickened edge construction joint

sawed or tooled control joints

d/4

d

half-round keyway or beveled 1"x2"

placed first   placed later   placed first

d/10 min

Keyway construction joint

reinforcement

2"x4"

2"x4" stakes, 4' on center, cut off flush

dirt that drains easily or 4" granular fill 1' minimum beyond edge

stop insulation 6" from top & bottom edges

#3 bars, 24" on center

masonry reinforcing, 24" on center, vertically

extend insulation to vertical edge

Joints, reinforcing wire mesh, and rebar all add structural integrity and reduce cracking and chipping in poured concrete.

strength slightly, but it greatly improves the ability of the concrete to stand up to freezing and thawing, salt, and other weathering processes.

Concrete blocks are suitable for laying up small foundations and walls. In these applications, they are mortared together, and they can be reinforced with rebar that runs up through the blocks.

ROOFING MATERIALS

Asphalt shingles and steel panels are probably the most common types of roofing because they are relatively inexpensive, reasonably good-looking, easy to install, and last for years. Asphalt shingles have varying degrees of fire resistance, depending on their design; steel roofing is fireproof. In areas subject to high snowfall, steel tends to allow snow to slide off the roof more easily than other types of roofing, but an avalanche of snow off the roof can scare animals and bury things (like hose bibs and tools) under a huge pile of snow. Other roofing systems, such as wooden shakes, tile, slate, and concrete, tend to be expensive. (See chapter 10 for more on roofing materials.)

Common Lumber Abbreviations

Abbr.	Meaning	Abbr.	Meaning	Abbr.	Meaning
AD	air-dried	FA	fascial area	RES	resawn
ALS	for softwood	FAC	factory	Rfg.	roofing
AV	average	FAS	firsts and seconds	RGH, Rgh.	rough
AW&L	all widths and lengths	Fb; Ft, Fc, Fv; Fcx	allowable stress (lb./sq. in.) in bending; tension, compression, and shear parallel to grain; and compression perpendicular to grain	R/L, RL	random lengths
BD	board			R/W, RW	random widths
BD FT	board feet			SDG	siding
BDL	bundle	FLG	flooring	S-DRY	surfaced dry; less than 19 percent moisture
BEV	bevel or beveled	FOK	free of knots		
B/L, BL	bill of lading	FRT	freight	SEL	Select grade
BM	board measure	G	girth	S-GRN	surfaced green; lumber unseasoned, greater than 19 percent moisture
B&S	beams and stringers	GM	grade marked		
BSND	bright sapwood, no defect	G/R	grooved roofing		
BTR	better	Hrt	heart	SQ	square
CC	cubical content	J&P	joists and planks	SRB	stress-rated board
CF	cost and freight	JTD	jointed	STD	standard
cft or cu. ft.	cubic foot	KD	kiln dried	STK	stock
C/L	carload	Lft, Lf	linear foot, linear feet	S1E	surfaced one edge
CLG	ceiling	LGR	longer	S2E	surfaced two edges
CLR	clear	LGTH	length	S1S	surfaced one side
Com	common	M	thousand	S2S	surfaced two sides
CS	caulking seam	MC, M.C.	moisture content	S4S	surfaced four sides
CSG	casing	MG	medium grain or mixed grain	TBR	timber
DIM	dimension	MLDG, Mldg.	molding	T&G	tongue and groove
DKG	decking	MS	mixed species	UTIL	utility
D&M	dressed and matched	MSR	machine stress rated	WHAD	worm holes a defect
E	edge	PAD	partially air-dried	WHND	worm holes no defect
EM	end matched	P&T	posts and timbers	WT	weight
				WTH	width

9. BASIC CONSTRUCTION

Keep It Square and Level

There is one very important thing to keep in mind: the quality of your final project will largely depend on your ability to keep things square and level. Take your time at each step, and constantly check your work for correct angles and elevations. You'll be happy you did when the project is done.

This chapter covers the basics of construction: foundations, framing, and roofing. Although plumbing, electrical, and finish work are covered in the next chapter, study up on those areas before constructing your foundation, because preparations for plumbing and electrical service may need to be done as the foundation is being prepared.

If you haven't already studied chapter 3, you might want to review the planning information on types of foundations, framing systems, and styles of rooflines now.

LAYOUT

Laying out a building is one of the most important steps in the construction process. Poor workmanship at this step dooms a project before the first nail is pounded.

The site must be prepared before you lay out the corners. For a small building on relatively flat and well-drained soil, site preparation might simply involve some shovel work to strip vegetation and topsoil (which should be piled away from the work area for use in final grading and landscaping) and to remove any large rocks or stumps. For bigger projects, it will require excavation with heavy equipment, and that's probably best left to a subcontractor with experience. See page 182 for an overview of how to level a transit, an important tool if you'll lay out the building yourself.

Locate Utilities

Before you do any excavation, even with a shovel, request that your local utility providers visit the site to locate all underground lines. After utilities are located, if you dig in an area the locator indicated was clear of utilities and cut a line, the utility provider is responsible for all costs associated with the cut; without a locate, you are responsible and, depending on the nature of the damage, the costs can quickly run into the tens of thousands of dollars. And if you accidentally dig up any kind of public utility (water or sewer lines, telephone or electrical lines, natural gas pipes), not only will the bill be hefty, but the accident might also result in serious injury or death.

Using a Transit or Laser Level

A transit, or theodolite, is a surveyor's tool designed to measure horizontal and vertical angles, thus allowing accurate layout. It requires two people — one to sight through the telescope and the other to mark points on a "story pole," which is a wooden pole that is marked for measurements, or to hold the leveling rod. Laser levels do the same work but can be operated by just one person.

A building can be laid out without a transit or laser level, but using one will make the work go faster, particularly for complicated structures or layouts requiring a great deal of excavation work. Both tools are somewhat expensive to purchase for a single project, but they can be rented at a tool rental shop.

There are three main parts to a transit: the telescope, the leveling vial or bubble, and the rotating plate, which allows the telescope to move 360 degrees and is supported on a tripod base. The key to successfully using a transit is getting the unit leveled initially. To level the unit, do the following.

1 Set up the tripod over the point you will work from using a plumb bob or preferably an "optical plummet," which comes standard on newer units, to ensure you are centered on the point. (For example, if you begin working from one corner of the building, be sure that the tripod is perfectly centered on the corner, with a nail or tack driven into a stake marking the exact center.) Get the head of the tripod level on the horizontal plane. On rough ground, set two legs to the same height and adjust the uphill leg to a height that provides a horizontal surface for the head.

2 When the tripod head is close to level, begin fine-tuning using the leveling knobs, which come in four-knob and three-knob patterns. When using the four-knob system, posi-

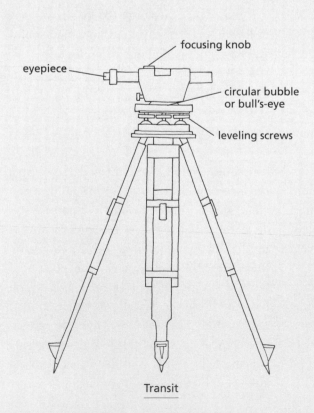

focusing knob

eyepiece

circular bubble or bull's-eye

leveling screws

Transit

Sighting through the telescope

tion the leveling vial over one pair of knobs and adjust them to center the bubble in the vial. Turn the telescope 90 degrees and repeat with the other set of knobs. For the three-knob system, use two knobs to center the bubble in one direction, then turn 90 degrees and use just a single knob to level in the other direction.

3 Use the "left thumb" rule. All students of surveying are taught this trick, and it makes leveling easier. Simply put, when both thumbs move in on the knobs or both thumbs move out on the knobs, the bubble follows the left thumb. So if both thumbs are moving out, the bubble goes left; if both thumbs are moving in, the bubble goes right.

4 After you think the unit is leveled, turn 180 degrees to check that you indeed have a level transit. If the bubble at 180 degrees is not centered, return to the first step and try again.

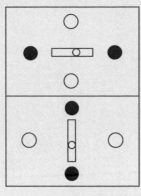

Four-knob system

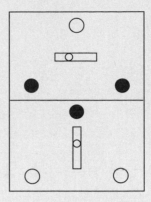

Three-knob system

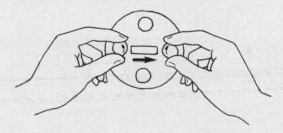

"Left thumb" rule: left thumb moves in, bubble moves to right

story pole

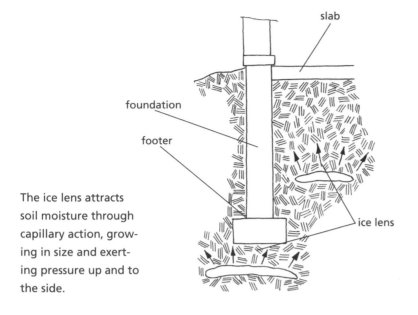

The ice lens attracts soil moisture through capillary action, growing in size and exerting pressure up and to the side.

Frost-Protected Shallow Foundations

The frost-protected shallow foundation has been used extensively in Scandinavian countries and tested by a U.S. Department of Housing and Urban Development project in cooperation with the National Home Builders Association. Basically, it combines a shallow concrete footing with rigid foam insulation placed around the foundation perimeter. The number of heating-degree days in your area is used to determine how much insulation should be applied.

The FPSF system has been successfully tested in the United States in places ranging from Minneapolis to the mountains of Colorado, and the American Society of Civil Engineers has published a user's guide. Building code organizations are in the process of adopting a standard for it into existing codes, so it is becoming a readily available alternative to deep footings. (For more information, see the resources.)

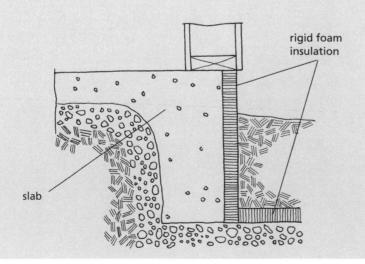

Many areas now have a "one-call" service; with this arrangement, you simply make one phone call and the company providing the service notifies all appropriate utility companies or subcontracts the locating for the utility companies. If your area doesn't offer this service, be sure to call every utility company that may have lines or pipes running anywhere on or near your property.

Excavating

The order in which you'll do the next couple of steps depends on the type of footings and foundation you'll be using. Small and shallow excavations are often easiest if you mark the exact corners first (as described below) and then remove material by hand. This works for small, dirt-floored sheds, like a chicken coop. Piers, pilings, and posts also require you to start by first marking the exact locations of the corners. For deeper footings and foundations, as required for a true basement or partially buried lower level, you will want to simply mark an approximate area that is wider than the area you will work in by at least 10 feet on all sides, and then precisely mark the corners when you have a suitable rough excavation for the hole.

Historically, for a footing/foundation system, you needed to dig deeply enough so the footing was below the frost depth in your area, but today you may be able to use a frost-protected shallow foundation (FPSF), which significantly reduces the required excavation depth in extreme cold-climate regions. The deep excavation was required to protect the footing/foundation and hold the building steady and straight in spite of the heaving action caused by soil ice (frost), which forms an ice lens. An ice lens grows over time because it continues to attract soil moisture, and as it grows, it exerts pressure that can push up a footing, thereby causing cracks in the concrete. But proper drainage and insulation have been incorporated into the FPSF, thereby offering the same protection without such a deep excavation.

Marking the Corners

Finding the exact locations of a building's corners and squaring a building take time, but aren't difficult to do right.

First, place stakes (1x2 stakes from the lumberyard work great for this step) at the approximate corners, measured off with a 50-foot or 100-foot steel tape stretched tight at each point. These stakes don't need to be buried deeply because you'll pull them out shortly.

Next, place batter boards outside these approximate corners of the building. (See the box to learn how to set up batter boards.) The deeper the excavation for the footing and foundation, the farther from the corner stakes you'll need to place the batter boards. For a shallow excavation or posts, set the batter boards 3 feet from your approximate corner

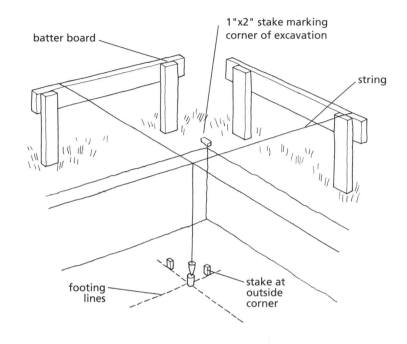

Batter boards and string are used to locate corners of buildings. A plumb line hanging down from the intersecting strings points to the exact location of the corner.

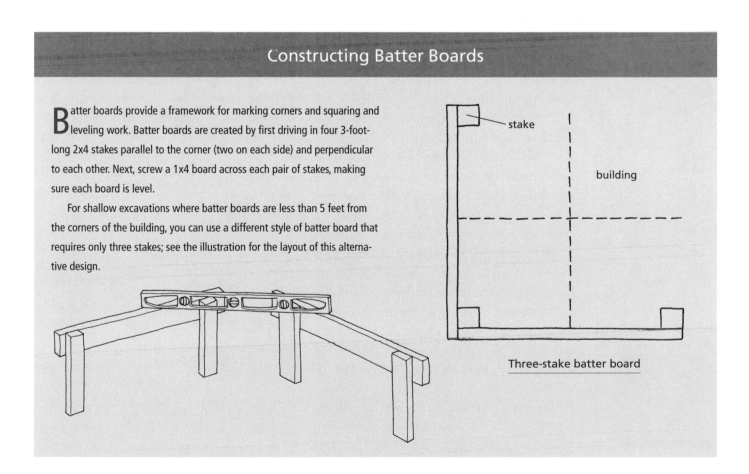

Constructing Batter Boards

Batter boards provide a framework for marking corners and squaring and leveling work. Batter boards are created by first driving in four 3-foot-long 2x4 stakes parallel to the corner (two on each side) and perpendicular to each other. Next, screw a 1x4 board across each pair of stakes, making sure each board is level.

For shallow excavations where batter boards are less than 5 feet from the corners of the building, you can use a different style of batter board that requires only three stakes; see the illustration for the layout of this alternative design.

Three-stake batter board

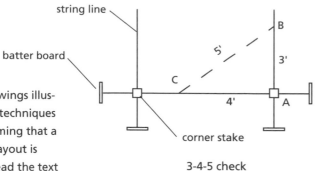

3-4-5 check

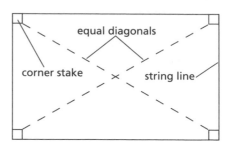

Diagonal check

These drawings illustrate two techniques for confirming that a building layout is square. Read the text for detailed instructions for the 3-4-5 check and the diagonal check.

TIPS

These tips will help speed things along.
• Have a helper hold the short side of a framing square against the board. When the string is flush with the long side of the framing square, it is close to where it needs to be.
• Make a model right triangle out of nailed-together 1x2s, with the measurements 6-8-10. It makes any corrections and checks quicker and easier.

stakes; for a deeper excavation, set them 10 feet to 20 feet outside the proposed corners.

The batter boards are laid out perpendicular to each other for each corner of the building; they should be at least 1 foot above the approximate height of your finished foundation, and level. Lightly tack a nail into the top edge of the batter boards facing each other on one long side, and tie a string between them so the string is over the approximate outside edge of the corner stakes you have already placed on one elevation of the building.

Now repeat the step so that strings intersect over all the approximate corners. The next step is to square the corners so the strings create perfect 90-degree angles, and this is easily done if you remember a little geometry. The 3-4-5 triangulation rule (a solution of the Pythagorean theorem that yields whole numbers) says that if side AB is 3 units long (or a multiple of 3) and side AC is 4 units long (or a multiple of 4), then side BC will be exactly 5 units long (or a multiple of 5). So set one of your 1x2 stakes with a finish nail centered under the point we are calling A at 3 feet, then go to the intersecting string and similarly mark B at 4 feet; draw your measuring tape between the two points. If you get 5 feet on that measurement, the corner is square; if not, take the tack out of the batter board for one of the strings and adjust it until you have a perfect 3-4-5 triangle.

Once you think you've got the strings squared well, there are two checks to do. First, expand your triangle by the greatest propor-

tions of 3-4-5 that you can. For example, if the building is 10 feet wide and 20 feet long, you can check at 9 feet and 12 feet, which should yield a 15-foot measure between points B and C. The second check is called the *diagonal check*, and it involves measuring from each corner on the diagonal: if the two diagonals are exactly equal lengths from corner to corner, you have a perfect rectangle; if not, you need to do some more tweaking.

The last step, once you have everything completely squared, is to secure the lines well in these positions. Retie each line to a nail securely driven into the back of the batter board, in exact line with the correct position. After the strings are tied tightly, check one last time to confirm they are still square. The intersections of these lines should mark the exact corners of your building.

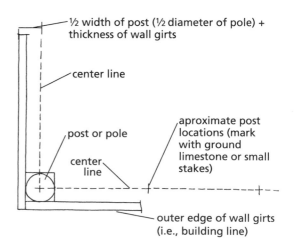

Mark the corners of the structure and the "centers" of every post or pole that will support it, because each support requires an excavation.

Piers, Pilings, or Posts

For construction using piers, pilings, or posts, you need to mark the exact centers of each of the planned holes (think the bull's-eye of each hole), including not only corners but interior supports as well. After the outside corners of the building have been marked with string, use 1x2 stakes to mark the exact centers of the holes you will need; these centers will be half the size of the footing, or pier, piling, or post if no footing will be included. After the first corner is marked, measure off and mark the exact locations of the rest of the posts required in the design. Once stakes marking the centers of all the excavation holes have been placed, you may need to move aside the lines from the batter boards so you can work. Finally, excavate the holes by hand using a shovel and posthole digger, or use a power auger, which can be either a handheld or a tractor-mounted unit. When excavating for piers, keep the hole as narrow as possible.

BUILDING A STRONG BASE

The footing/foundation system is designed to accomplish a couple of things: it disperses the weight of the structure over a large area of soil and it secures the building to the soil so it can't be dragged or blown away. Footings are always created with poured concrete. Foundations can be poured concrete, concrete block, or treated wood.

In poorly drained soils, it's a good idea to provide drainage around the footings, which increases stability, particularly in areas where freezing is possible. In extremely wet areas, use a drainpipe to divert water away from the footing; in moderately wet soils, simply backfill the trench with screened rock (¾ inch to 1½ inches), river cobble (2 inches to 6 inches), or concrete rubble.

Loads

In chapter 3, you read about the load-bearing capacity of different types of soil. This information is critical when sizing footings and foundations. For design purposes, the bearing capacities typically associated with different soils are shown in the chart below.

To think in simple terms about what different soil types mean to a footing/foundation system, consider this: a building constructed on soft clay requires four times the area of footing (in square feet) than a building built on hard rock.

Generally, if a permit is required to construct a building, the building official will also

LOAD-BEARING CAPACITIES OF SOIL	
Soil Type	Load capacity (lbs./sq. ft.)
Soft clay, sandy loam, or silt	1000
Ordinary clay or sand	1500
Sandy gravel, gravel, sedimentary rock	2000
Massive crystalline bedrock	4000

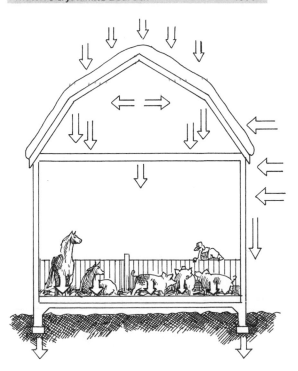

Many types of loads have to be considered when constructing a building. *Dead loads* are the weight of the structure and permanent fixtures that are incorporated into it; *live loads* are the weight of animals, people, equipment, and snow, as well as pressure applied by wind. All impact building design and performance.

require that a soil analysis be conducted by a registered engineer to determine its load-bearing capacity and the depth to the water table if you are located in an area with shallow water tables. The engineer will then prepare recommendations for the foundation type and design criteria. Even if you aren't required to get a permit, it is nonetheless important to ensure the footing/foundation design is adequate to support the building under the types of expected loads. If you are good with math and understand some basic engineering principles, you can get a copy of the code and do the design yourself. If not, talk to your local building official (he or she will have knowledge of local conditions and the standards generally applied by the code), or check with your Cooperative Extension Service office, which might be able to schedule a visit from an Extension agricultural engineer, who can help you with the design.

Lateral Stress

Most often, vertical stress (or the stress that pushes straight down on the foundation from the live and dead loads) is the biggest issue when considering load, but in some cases a building creates a significant horizontal, or lateral, stress. For example, arched buildings tend to push out on the foundation horizontally. Lateral forces also become an issue in basements with bulk grain stored against their walls. When lateral stress is a concern, the foundation needs to be beefed up with buttressing. Foundations that will have this type of lateral stress should be designed by an engineer.

Compaction

Soil that the footing/foundation system will rest on should be compacted to reduce movement. For a small project in relatively stable soils (the higher the bearing capacity, the higher the stability), you may simply need to do some hand tamping within the footing area itself with the butt end of a 2x4. If you will be pouring a slab or building a large structure on less stable soils, you should provide additional compaction. For this bigger job, a gasoline-powered hand tamper, available at tool rental stores and some hardware stores, makes the work quicker and easier and produces a better result. For an expensive structure that's going to be built on highly unstable soils, you may need to overexcavate by as much as 4 to 5 feet and truck in road base that will be placed and compacted in 1-foot lifts until the proper

---

Example Calculations for Footer Size Based on Building Loads

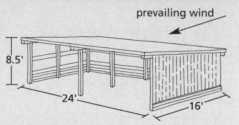

prevailing wind

Let's walk through an example of calculating the impacts of building loads on footer size for a simple loafing shed. We'll assume we are in an area with snow loads of 30 pounds per square foot and that the roof system and side wall dead-load at 6 pounds per square foot (3 pounds per square foot for the corrugated metal sheeting and the same for the wooden system of rafters, purlins, and nailers). Because it is a dirt-floored structure, there is no live load from the weight of animals or equipment parked in it. Remember, the load is calculated by the following formulas:

(live load + dead load) x (area) = load in pounds

Roof: (30 lbs./sq. ft. + 6 lbs./sq. ft.) x (16 ft. x 24 ft.) = load in pounds
36 lbs./sq. ft. x 384 sq. ft. = 13,824 lbs.

Walls: (calculate only for the bearing walls; in this case the two side walls)
(0 lb./sq. ft. + 6 lbs./sq. ft.) x (8.5 ft x 16 ft. x 2) = load in pounds
6 lbs./sq. ft. x 272 sq. ft. = 1632 lbs.
Total load = 13,824 lbs. + 1632 lbs. = 15,456 lbs.
Load per column = 15,456 lbs. ÷ 9 columns = 1717 lbs.

Now look what happens to this building under two different soil conditions: clay soil with a bearing capacity of 1000 pounds per square foot and gravel with a bearing capacity of 2000 pounds per square foot.

1717 lbs. ÷ 1000 lbs./sq. ft. = 1.7 sq. ft. of footer per column (approximately 15-inch diameter) or
1717 lbs ÷ 2000 lbs./sq. ft. = 0.85 sq. ft. (approximately 11-inch diameter)

grade is reached. *Road base* is a mixture of dirt and gravel that's prepared with compaction in mind. If the soil is unstable, be sure to work on the design with an engineer.

Utilities

Plan ahead if you will be incorporating any underground utilities (water, sewer, electric, natural gas, propane) in your project. They may need to come in under the footing or through the foundation wall or slab. Thinking about them before you pour concrete will definitely save you money and headaches later.

Preparing for utilities before pouring concrete is called *roughing in*, which consists of setting pipes or conduits so concrete can be poured around them. Utilities may require additional excavation, and many of the same concerns with compaction apply to these excavations as well as to the foundation. Water and sewer lines should be well bedded on a compacted base and below the frost line.

Forming Footings

Continuous footings for wall foundations may be formed in a trench dug below grade or by using forms above grade. When constructing wall foundations, the foundation wall must be a little wider than the wall it will support, and the footing is usually twice as wide as the foundation wall. For example, if you will be constructing a wall using 2x6 lumber, the foundation wall will be at least 8 inches and the footings at least 16 inches. Also, for a continuous wall foundation, the footing should include a "key," or shallow channel in the footing, that allows the foundation wall to lock into the footings. The key can be formed by stabbing a beveled 2x4 into the floated concrete, but it should be left partially above the surface of the pour.

Piers, posts, and pilings use a pedestal type of footing. Footers for these should be at least 3 feet deep to provide lateral support for the wall system against wind stress.

Slabs are special in that they may be built on a prepoured footing, poured on gravel between basement walls, or poured as footing and slab in one unit (these are known as monolithic pours, or floating slabs, and are discussed later). The advantage of the prepoured footing is that you will spread work out into separate tasks, which is sometimes helpful if you will be doing all the work yourself. But if you are hiring a concrete contractor, he'll generally opt for the monolithic pour, unless he's precluded from doing so by code.

All footings should be level and square, and they should include rebar for structural integrity. Rebar should be continuous with bends at corners, should overlap where two pieces meet by at least forty times the diameter (e.g., if using ½-inch rebar, the overlap would be 20 inches), and should be at least 2 inches above the ground within the pour. The rebar can be held in position using rocks or wire "chairs," which are specially designed for the purpose.

Footings can be stepped to accommodate grade differences. This technique is easier when you're planning to use concrete blocks than with poured walls, but it can be done with either design.

Rubble Foundations for Straw-Bale Structures

Remember the rule that says the foundation wall needs to be wider than the wall that will sit on top of it? Well, when using bale-construction techniques, that could make for a darn wide footer and foundation wall! But there is an alternative for this situation: a rubble trench foundation (RTF). If using this system, dig a trench at least twice as wide as the bale wall in clay soils — and that's just slightly wider than the bale wall in gravel soils — to below the expected frost depth for your area. Install a drainpipe at the bottom of the trench, then fill the trench with rubble (washed stone ranging from ¾ inch to 3 inches works well). Compact the rubble as you fill the trench and top it off with a layer of landscape fabric to prevent the soil from filling the gaps in the rubble. If the gaps were to fill, the rubble base would be subject to heaving when frozen, but by keeping soil out of the rubble, it drains and doesn't heave.

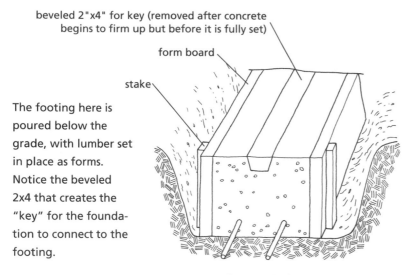

beveled 2"x4" for key (removed after concrete begins to firm up but before it is fully set)

form board

stake

The footing here is poured below the grade, with lumber set in place as forms. Notice the beveled 2x4 that creates the "key" for the foundation to connect to the footing.

Below-grade footing

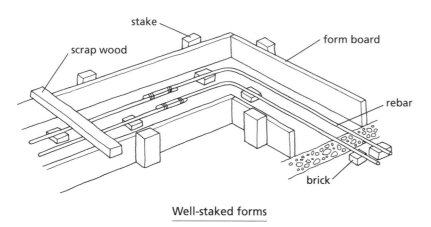

stake

scrap wood

form board

rebar

brick

Well-staked forms

Forms must be well staked so that they don't "blow out" when the concrete is poured into them. This illustration also shows how rebar is set within the forms.

In a trench footing, the trench acts as the form. In stable soils where it's possible to create a clean trench, this technique works well, but in some soils it is almost impossible to prepare a good trench. The 2x4 for the key has already been pulled from this pour.

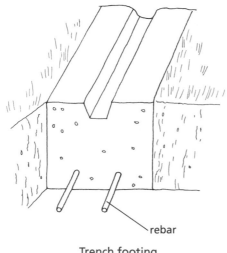

rebar

Trench footing

Anchors and Fasteners

You'll need to set anchor bolts or specialized fasteners that connect foundation walls or posts to the footings and that connect final walls to the foundation. There are two approaches: wire them into position prior to the pour or stab them into the concrete after the pour is complete but before it has begun to set up. Anchor bolts connecting a continuous wall foundation are typically set within 1 foot of every door and corner and at 6-foot intervals between; 6-inch-long J-bolts work well.

When a footing is being prepared to support a concrete-block wall, insert pieces of rebar instead of bolts. The rebar is generally cut so it goes through all the rows of blocks, and so that every fourth block has a piece of rebar extending to just below the surface. Fill the cores of these block columns with concrete, and before the concrete dries, add J-bolt anchors. For added insulation, the other columns can be filled with foam insulation or vermiculite.

Working with Concrete

When your forms are ready, you can pour the concrete. How much concrete will you need, and how do you get it? Well, that depends on the size of the pour (see box).

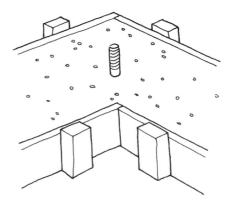

After concrete is poured and the surface is smooth, "punch" the anchors and fasteners into it. Forms must stay up until concrete is fully set, not until it is fully cured. Depending on the size of the pour, they can be pulled in 1 to 5 days.

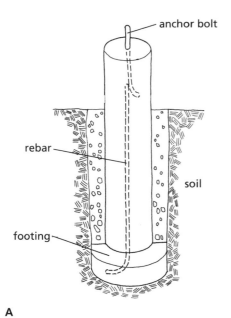

A

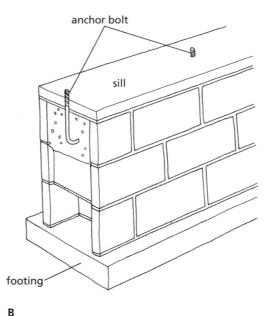

B

Rebar and anchor bolts (or connectors) need to be included in pours for stability and for connecting framing to foundations. Illustration A shows rebar and an anchor bolt in a poured pier; illustration B shows J-hook-style anchor bolts in a concrete block wall.

Although it's theoretically possible to pour any job using dry premix concrete in bags, it would be very challenging to do the job in the first example (see box at right) that way. You would need about 120 bags, which are heavy and hard to work with. And when you begin a pour, as for continuous footings or a slab, you need to do the entire pour at once. So, contracting with a ready-mix operator to bring the concrete to your site and do the pour quickly, efficiently, and with minimal manual labor would be best for this job. Bags of premix would probably be fine for the job in the second example. Each column would take about a bag and half, and it wouldn't be necessary to do all nine pours on the same day.

Another term you'll hear when talking to a concrete supplier is *slump*. Slump is a measure of the consistency of the concrete, which is affected by the amount of water it contains. Higher slump means more water in the mix, which means it's easier to work. But higher slump also means there will be more shrinkage and cracking as the concrete dries, and that may reduce its structural integrity. Work with as stiff a mix as you can — a slump of 3 to 4 is good for walls when working by hand, and 4½ or greater works well for flat work.

How Much Is Enough

Concrete is purchased by the cubic yard, or as contractors simply say, "by the yard." One cubic yard is equal to 27 cubic feet. So the first thing to do is calculate the volume in cubic feet by multiplying the length by the width by the depth of the pour (for round pours, use the equation πr2 x depth, which equals 3.1416 x ½ the diameter x ½ the diameter x the depth). Here are a couple of examples.

Example 1: Assume you are pouring a footing for a wall foundation under a square building that measures 15 feet on each side. The footing is 16 inches wide and 8 inches deep. The total length will be 60 feet (15 feet per side x four sides). First, convert the measurements in inches to feet so you are working with the same units:

16 inches ÷ 12 inches per foot = 1.33 feet wide;
8 inches ÷ 12 inches per foot = 0.75 feet deep

Now find the total cubic feet of the pour:

1.33 x 0.75 x 60 = 60 cubic feet

If you will order concrete from a batch plant, you'll need to convert the total cubic feet to yards. Do this by dividing the cubic feet by 27.

60 ÷ 27 = 2.22 yards

Example 2: Assume you are pouring footings for the nine columns that hold up the shed in the box on page 188. The footings should be 15 inches in diameter and 8 inches in depth. Again, convert to feet (15 ÷ 12 = 1.25 and 8 ÷ 12 = 0.67) and then use the equation 3.1416 x half the diameter x half the diameter x the depth:

3.1416 x ½ x 1.25 x ½ x 1.25 x 0.67 = 0.82 cubic feet per column.

Now find the total yards:

0.82 cubic feet/column x 9 columns ÷ 27 = 0.27 yards

spade to eliminate voids and honeycombs

straight board to level concrete

stake cut flush

dampen soil before pouring concrete

prepared subgrade

float to smooth; wait until bled water has evaporated

Tools for pouring concrete include a spade or trowel for eliminating air spaces and honeycombing; a float, which can be a 1x6 board; and a leveling device, which can simply be a straight board. Moisten the soil before pouring the concrete.

caution

For small pours, one or two people can easily do the work, but if you will be undertaking a large pour, recruit some helpers. Be sure to remind your helpers that concrete can burn, so rubber boots and gloves must be worn, particularly if anyone will walk in the concrete; this usually isn't much of a problem when doing footings, but it becomes a major concern when pouring big slabs.

TIP

The minimum depth for a footing is 8 inches in no-frost areas.

If you are mixing small batches of concrete in a wheelbarrow, place the dry mix in the wheelbarrow, add water gradually, and mix with a garden hoe or spade. Drag the hoe or spade through the mix to make a trough. If the sides of the trough are crumbly and the concrete falls in chunks when you disturb it, then you need to add water (one cup at a time), mixing after each addition. If the mix is soupy and runs back into the trough, add more dry mix. The mix is at proper slump when the ingredients are thoroughly wet, the sides of the trough stand, and the concrete has a shiny surface when patted by hoe or spade.

Dampen the earth and forms with a garden hose before starting the pour and then fill the forms to the top. As you pour, use a shovel to work out air pockets in the concrete. Using a shovel or a trowel, make sure the concrete fills corners firmly. Next, use a metal or wooden float (which on small pours can simply be a piece of 2x4) to "float," or smooth out, the top of the pour. At times you may need to dampen the surface of the pour to float it; a light misting from a garden hose will usually do the trick. After the concrete is smoothed but before it begins setting up, place anchors.

Finally, to minimize cracking, immediately tool joints into the concrete surface before it sets, and to improve traction set grooves or roughen the surface of concrete slabs on which animals will walk.

Remember that once you begin pouring, you need to work quickly. Concrete that's allowed to begin drying as more is being poured forms a "cold joint," which is a structurally weak area.

Sometimes it becomes necessary to pour new concrete adjacent to concrete that has already set (for example, you might add a slab after the fact to a building that originally had a concrete foundation and a dirt floor). When this is the case, install an isolation joint. You can create an isolation joint by inserting an asphalt-impregnated expansion strip or a piece of 1-inch lumber coated with asphalt roof coating between the existing concrete and the new pour.

Remember, concrete needs to cure, which can take anywhere from several days to several months, depending on the size of the pour and the ambient temperature. The optimum is 28 days, but you can usually get back to work within 4 to 7 days after the pour.

The Foundation

When the footings have cured well (1 to 4 days, depending on the weather), you can begin work on the foundation. It pays to reset the guide strings from your batter boards at this point so the strings clearly mark the outside corners of the walls themselves. This helps both in building forms for poured walls and for laying up blocks straight and true.

Foundation walls should extend above ground level. For a wood-sided structure, plan for a foundation wall that is at least 12 inches above ground level; 6 inches is sufficient for a building that will be sided in steel.

Construct solid forms for the foundation pour. They will be holding far more concrete than the footings, and the weight of the concrete can easily split apart forms that are not well supported, leaving a mess that's hard to clean up, not to mention a waste of money.

Slabs

Slabs are usually poured 4 to 6 inches thick on a bed of gravel or sand that has been smoothed within the forms. Vapor barriers can be added under the layer of gravel and insulation on top of the gravel layer to help keep the slab warm and safe from frost heaves.

The surface of the slab should be poured about 4 inches higher than the outside ground level; this is low enough to enter without a step by just sloping some dirt up to doors but high enough to allow drainage away from the structure. Slabs are reinforced with a combination of rebar and wire mesh.

Pier and Post Foundations

Pier and post foundations are common for barns. Although they can be square, the introduction of fiber forms has made round pours the most common for these types of foundations. The forms are inexpensive and easy to use. The form should be set so that it is at least 4 inches higher than the surrounding ground level, but 6 inches is better still. In post construction, the post may be set in concrete all the way to the top, or, if posts are pressure-treated, partially formed in a concrete collar and then backfilled with well-packed aggregate, such as road base, that's hand tamped in layers as it is brought up to the surface. Don't use native soil for backfilling posts.

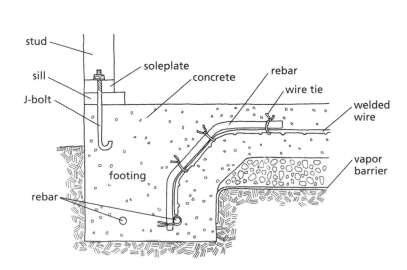

Monolithic slab

These are the elements of a monolithic slab. Slabs require welded-wire reinforcement throughout the pour as well as rebar.

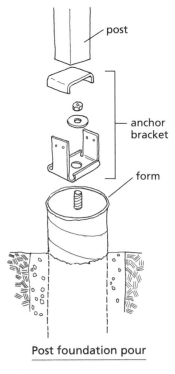

Post foundation pour

The typical elements of a post-building foundation pour are shown here. The concrete is poured in a fiber form. A J-hook anchor bolt is set and then the anchor bracket included on top. The post connects to the bracket.

Slabs and foundations can be insulated with rigid foam to keep floors warmer and to reduce the impact of frost. When including insulation, protect it from the weather where it extends above ground with a drip edge and an exterior plywood face.

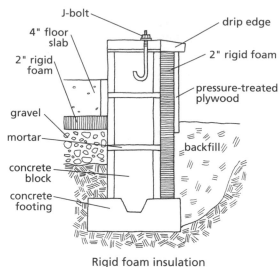

Rigid foam insulation

The weight of wet concrete can split apart poorly braced forms, so be sure the forms for poured walls are well braced.

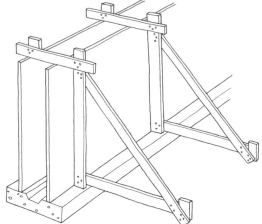

Well-braced forms

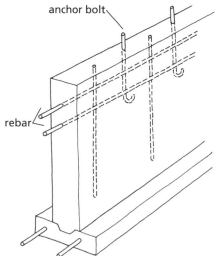

Rebar and anchors

This is how rebar and anchors are typically used in a wall foundation. The J-hook catches the lower piece of rebar and is wired to it for extra stability.

Wall Foundations

Two-story barns usually use a wall foundation because of the size and weight of the structure. Some barns include housing for guests or helpers, and these also tend to incorporate a wall foundation. Short wall foundations (sometimes referred to as *trench foundations*) are cost-effective in areas not subject to freezing because they don't have to be very deep; in cold climates, wall foundations can easily be insulated to help maintain building temperatures.

In some cases, a combination wall-and-pier system makes sense. For example, when building into a hillside, a wall is used for the hill portion and piers for the part away from the hill.

Partition walls that are load-bearing need to be supported. This can be done with interior wall foundations or, more commonly, piers.

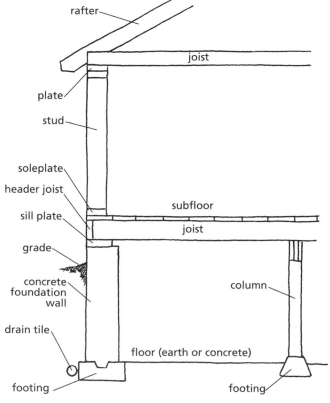

Major structural features

FRAMING IT UP

When the foundation system has been completed, you are ready to begin building the frame, which supports the building's loads and transfers them to the ground via the foundation. We are beginning to see more nonwood alternatives in the construction industry, from light-gauge steel framing to insulated concrete forms to prefabricated plywood panels, but wood is still the most common framing material for most owner-constructed, smaller animal housing projects. There are several types of wooden framing systems.

• **Stud-wall framing.** Stick-built construction is the most common technique for framing small structures. There are two approaches to stud-wall framing:

—**Balloon framing.** Balloon framing has continuous studs rising directly from the foundation to the eave and is built in place instead of on the ground. The floor is constructed after the exterior walls are up. Balloon framing has fallen out of favor because it takes longer to construct, won't carry as great a second-story load, and tends to burn faster than a platform-framed structure; though when all things are equal, it is less vulnerable to strong lateral winds than a multistory platform. (One exception to the caveat regarding second-story loads is when balloon framing is used with timber-frame construction, which can effectively carry a tremendous second-story load.) The technique is still used regularly in creating stairwells and areas that are open up to high ceilings.

—**Platform framing.** Platform framing is done on the ground by constructing the walls as a series of panels that are then raised into position. As each panel is raised, it's braced into place until it can be tied into the other walls. Long walls can be laid out in multiple short panels. Currently, platform framing is the most com-

mon method of building a stick-built structure. For multistory structures, each floor is constructed as an independent unit that is stacked on top of the lower unit. Platform framing provides a sturdy construction design.

• **Timber framing.** Also called post-and-beam framing, this technique uses age-old methods of joining large, structural timbers with dovetails and mortise-and-tenon joinery instead of nails or other metal fasteners. (If you are seriously considering constructing a timber-frame barn, see the resources for books on the topic.)

• **Post framing.** Post framing is similar to timber framing in that large structural members support walls, but it uses metal fasteners. In spite of its similarity to timber framing, it is generally reserved for single-story structures.

Stud-Wall Framing

Studs are the vertical boards that make up the frame. Two-by-fours are commonly used for single-story buildings and 2x6s for two-story structures, though the extra expense for 2x6 lumber is fairly insignificant, so you may opt

Post-framed buildings that will be covered in metal use girts and purlins as nailers for the siding and roofing, respectively, and to add stability to the structure.

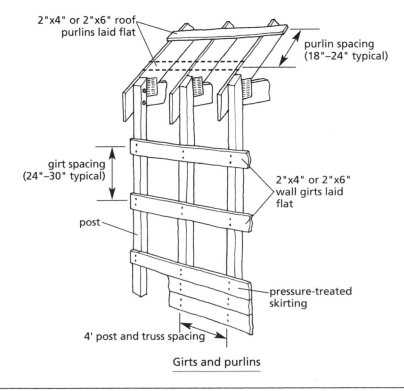

2"x4" or 2"x6" roof purlins laid flat

purlin spacing (18"–24" typical)

girt spacing (24"–30" typical)

2"x4" or 2"x6" wall girts laid flat

post

pressure-treated skirting

4' post and truss spacing

Girts and purlins

TIP

Center refers (when saying, for example, that studs are laid out on 24-inch centers) to the distance from the middle, or center, of one board to the center of the next board.

for 2x6s if you want to insulate a building well or expect heavy loads; 2x4s or 2x3s can be used for interior, nonbearing partition walls. (Two-by-eights or offset, double rows of 2x4s are now finding their way into residential construction because they allow for the use of extra insulation and so provide significant energy savings over the life of the house.)

Although 2x4s and 2x6s are available in lengths ranging from 8 feet to 18 feet, if you purchase lumber marked as "studs," they will be precut in lengths that yield 4-foot to 10-foot walls. What I mean by *yield* is that the stud, when nailed to the plates (two on the top and one on the bottom), yields a given length of wall. For example, an 8-foot stud is not 96 inches tall; it's actually 91½ inches long, but when it is nailed to the plates, the wall is 96 inches, or 8 feet, long.

Studs may be laid out on 12-inch, 16-inch, or 24-inch centers, depending on the stud dimensions (2x4 vs. 2x6) and the structure's height. Stud walls are connected to the foundation by a sill when a continuous wall foundation is used, or by a beam when piers are used.

The sill is set onto the anchor bolts you installed before the concrete dried and is secured using ½-inch nuts and washers. Some large projects call for a double sill (two pieces of lumber stacked with the edges set so that their seams are offset).

The position of the sill on the foundation depends on the type of sheathing system you'll use. In buildings that include a solid floor, the subfloor (also called the *deck*) is built on top of the sill and below the stud wall; in buildings with dirt floors, the stud wall is built directly on the sill. It's a good idea to use pressure-treated lumber for sills and to set them on a gasket that protects the sill from moisture and reduces drafts. In areas where termites are a concern, a thin metal termite shield is placed under the sill.

When laying out a stud wall, choose the straightest boards you can find for the soleplates (bottom plates) and top plates; crooked plates require more work to plumb the walls. Using a carpenter's pencil, carefully mark the soleplate and the first top plate (or framer's plate) for the position of each stud (including the cripple and jack studs that will be required for window and door openings) before you begin nailing the studs into place.

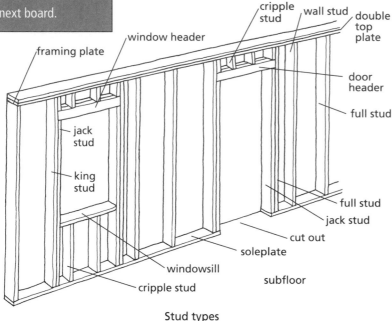

Stud types

Full studs are the primary wall studs that run from the soleplate to the top plate, though when they are adjacent to a door or window opening, they are referred to as *king studs*. Jack studs run between headers and the soleplate. Cripple studs run between the top plate and headers or between the sills and the soleplate.

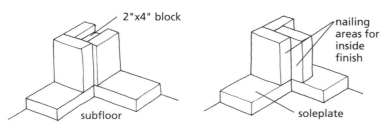

Corner construction

These are three different approaches to creating corners in stud-wall construction. Each provides a nailing surface, or method to attach drywall or other inside finishes, at the corners.

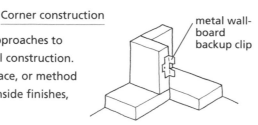

Framing a Wooden Floor

Wooden floors sit on a box sill. The box sill consists of headers, joists, bridging (or bracing), and possibly girders, supported by piers or an internal wall foundation if the span is too great for a single piece of lumber to cover the distance.

The span a joist can cover varies based on the live load the floor will have to carry, the type of wood, the spacing between the joists, and the size and quality of the board used for the joist. For example, when a 40-pound live load is expected, a select structural 2x6 of spruce/pine/fir can span 15 feet on 12-inch centers but only 11 feet 11 inches on 24-inch centers. If the same spruce/pine/fir 2x6 is purchased in No. 3 grade lumber, it will span only 11 feet 3 inches on 12-inch centers and 8 feet on 24-inch centers.

Joists can actually hang from the header in metal joist hangers, which are easy to use. Bridging helps steady the floor and disperse the load; it consists of either metal braces or 1x3 pieces of wood nailed crosswise between the joists. For wide spans, install a girder supported by a pier system. There are two approaches to using girders: In one, the foundation is designed to include cut-out "girder pockets," so the girder is below the elevation of the joists, which rest on top of the girder. The other approach is to treat the girder like a header and hang the joists off it as you would hang them off the headers. Girders may be a solid beam of wood, a built-up beam of wood constructed from two or three pieces of lumber with offset joints, or a steel I-beam.

Once the joists are all in place, it's time to set the subfloor, or deck. The most common decking material for any kind of animal housing is ¾-inch exterior-grade plywood. Because it installs quickly and easily and tolerates moisture and temperature swings better than wood composites like wafer board, it will usually be

your finished floor. But here is one place where Ken and I have a real preference for the not so common: we like to use 1x10 boards of native lumber for decking in all our outbuildings. The ever-so-small gaps between the boards allow moisture to drain out, the floors are sturdy, and native lumber is more attractive as a finished floor than plywood. Although you can nail down decking, wood screws work better for snugging the decking to the joists, and they tend not to work themselves out over time, as nails often do.

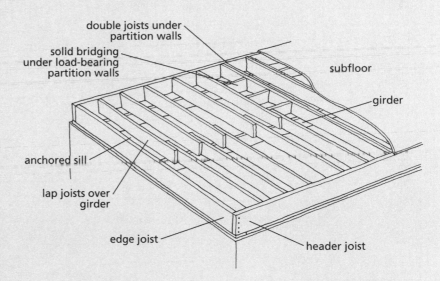

double joists under partition walls

solid bridging under load-bearing partition walls

subfloor

girder

anchored sill

lap joists over girder

edge joist

header joist

Joists, which may be referred to as *stringers,* are the support system for the floor. Joists overlap at girders set in girder pockets.

TIP

The best way to design a joist system is to contact an engineer (your local Extension agent may be able to get you engineering assistance from an agricultural engineer), architect, or the local building official.

TIP

Openings for doors
and windows need
to be larger than the
size of the door or
window to allow for
the jamb and shims.
Typically, openings
should be 1¼ inches
bigger on each side
than the actual
measurements of
the door or window.

It's quickest to mark both plates at the same time; lightly tack them together, so the marks will be identical on each. Run marks down both sides where each stud will be placed. Pull the plates apart and mark on the wide plane. Note the type of stud at each mark: *x* for full studs, *c* for cripple studs, *j* for jack studs, or *b* for blocks that are incorporated into corners for nailers.

Next, cut boards for headers, cripple studs, or jack studs. Work from one end to the other along the soleplate, nailing all the studs into place, and then work from one end to the other along the framing plate, nailing in the full-length studs. Finally, nail in headers, sills, jack, and cripple studs for doors or windows. Headers are constructed by nailing two pieces of lumber together, with blocks sandwiched between.

Studs are nailed with two 16d nails in each position — for example, two nails through the soleplate into each stud and two nails through the top plate into each stud.

Mark the floor (or sill) with a chalk line for the placement of the soleplate. Have bracing ready before you erect the wall and brace generously. For large walls, 1x4s can be screwed in across the face of the wall to help keep it straight, and cleats can be used for bracing to the floor and wall. Screw temporary floor stops along the outside of the building to help brace the wall when you raise it into position. Raising the wall definitely requires helpers, and for this phase, the more the merrier. When the wall is in the correct position, nail the soleplate to the sill, beam, or subfloor, and continue with the next wall. The top plate should cross all seams of the framer's plate, thereby enhancing the structural integrity of the walls.

Partition walls come after the exterior walls are done and are added in much the same fashion. Studs for partitions are set on 24-inch centers. Where a partition wall joins an exterior wall, the top plate comes all the way through to the exterior edge.

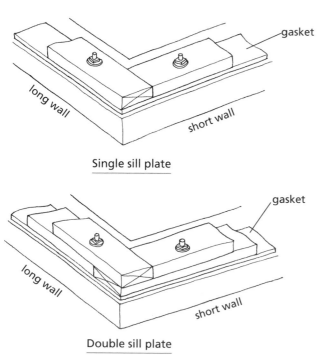

Single sill plate

Double sill plate

Sill plates can be installed as a single plate or a double plate, depending on structure size. Place a layer of gasket material under the sill to reduce moisture damage, and in termite areas, add a metal "termite shield."

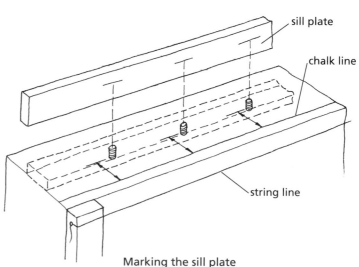

Marking the sill plate

Mark the sill plate so the holes will slip over the anchor bolts. Run a straight chalk line down the concrete before marking the board, and the sill will be straight and true when it is dropped over the bolts.

RAISING AND BRACING WALLS

To raise a wall requires helpers, but before you heft it up, screw temporary floor stops to the outside edge to help steady it.

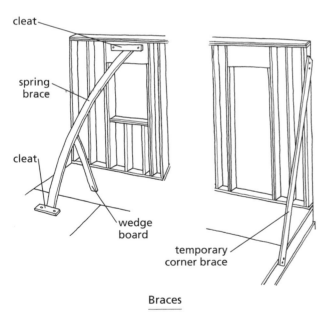

Braces

A spring brace or corner brace can be used to steady the wall after it is lifted into place. Use double-headed nails on cleats for easy removal.

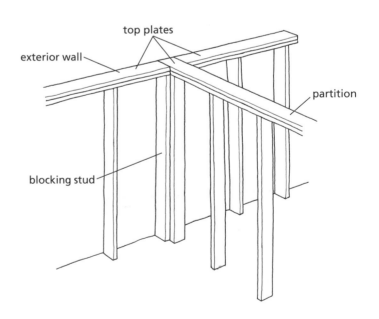

Where a partition wall joins an exterior wall, extend the top plate of the partition over the framing plate of the exterior wall and butt the top plates of the exterior wall snug against it.

Rib Construction

For really small sheds less than 10 feet wide, you may opt for rib construction. Ribs are individual framing units created using gussets (plates used to join two pieces of wood in the same plane and cut from plywood or purchased as metal units). The ribs are ultimately connected together by the rim joist, the roof and wall sheathing, and the subfloor. Additional 1x4s may also be added for bracing. Rib structures are often built on skids to raise them off the ground so they stay dry and to facilitate moving them.

To construct a rib structure, first build all the ribs you'll need to span the length of the building. As with other types of construction, have temporary bracing materials ready as you set the ribs into position.

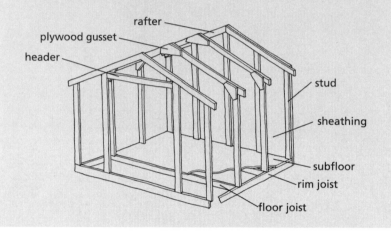

rafter
plywood gusset
header
stud
sheathing
subfloor
rim joist
floor joist

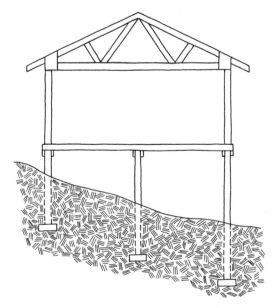

Although post-frame buildings are usually one story and built on a slab or dirt floor, they can have an elevated floor for second stories or sloping sites.

TIP

An agricultural engineer from the Cooperative Extension Service can help you determine appropriate sizes and spacing of posts, and truss design for conditions in your area.

Post Framing

Post construction is probably the most popular approach to building larger barns and stables because it's economical and relatively easy. It can be done using trimmed square posts, trimmed round posts, or with naturally tapered round posts, any of which should be pressure-treated to reduce rot. Naturally tapered posts (like telephone poles) have a slightly wider base than top, making them structurally stronger but a little more difficult to size evenly. Post-frame structures can incorporate "suspended floors" for second stories or on sloping sites.

The size and spacing of posts depend on the spans and the loads the building will have to sustain (4-inch to 10-inch posts are the norm). Eight-foot spacing is probably most common, but some structures may need 4-foot or 6-foot spacing, depending on the type of wood, the height of the eave, the size of the post, and the anticipated loads created by the wind. Although some designs have rafters or trusses spaced identically to the posts, most call for rafters or trusses to be added between the posts. (*Trusses* are specially engineered rafters composed of a series of triangles that allow for greater spans; see next page for more information on trusses.) In the designs that call for additional rafters or trusses, a double-plated girder is added to hold the rafters or trusses and help spread the load.

Each post needs to be temporarily braced as it is installed. When all the posts are in place and correctly leveled and squared, add the girts, or boards run between the posts, to the outside. Girts provide a surface for nailing siding and additional permanent bracing, thereby helping to stabilize the structure before girders, rafters, or roof trusses are added. Additional bracing, including purlins, may be called for before siding and roof sheathing are applied. Raising posts and trusses for all but the smallest buildings is easy if you have a tractor with a bucket on it for medium-sized buildings or a crane (available from rental services) for large buildings.

ROOFING

Like everything else, rafters need to be able to support the load that will be placed on them. Factors affecting them include the live and dead loads, the size of lumber used, the span, the angle of the roof, or pitch, and the type of wood used.

Trusses are a specially designed rafter system capable of spanning large areas and are common in larger barns. They are engineered using a series of upper and lower *chords* and *webs* that are held together by gussets. The webs create a series of triangles, which inherently add strength to the structure. They eliminate or minimize the need for inner supports (such as internal posts or load-bearing interior walls). For large projects, webs are generally purchased as prefabricated units from a local

building supplier, but you can build your own, particularly for smaller projects; they are easy to build on the ground.

Shed roofs, which have only one plane, or slope, are the easiest to build. The rafter is set to ride on the top plate of the sidewall in a notch called a bird's-mouth cut (see illustration on page 204).

Gable roofs are slightly more complicated to construct, but they aren't too difficult. The rafters connect to a ridge beam and are set on the side wall on a bird's-mouth cut. Many other roofs, such as the saltbox, gable and valley, half-monitor, and monitor, are a combination of different gable configurations. Gambrel and gothic-arch roofs are the archetypal roofs seen on traditional barns throughout the United States. They definitely add complexity to a roofing project. Barns are also sometimes built with hip, mansard, or conical roofs, and these too are more complex.

TIP

If you will sheathe or line exterior walls with plywood or oriented strandboard, do so *before* setting roof rafters or trusses. (See chapter 10 for more information on sheathing.)

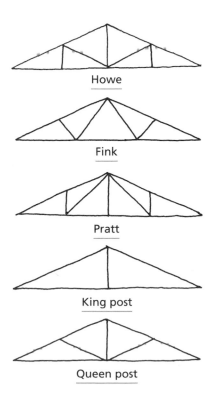

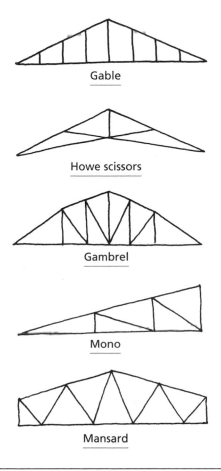

Trusses can span greater distances than rafters and can be used instead of rafters. Some of the more common rafter designs are shown here.

CAUTION

Installing roofs is dangerous — period. Jobs that are done high above the ground are best done on a secure working platform: either build a platform from wood or rent scaffolding. For safety reasons, on all but the smallest jobs, roofing is really a two-person (or more) operation.

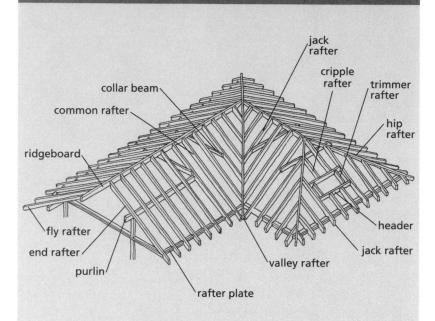

Roofing Terms

jack rafter
cripple rafter
trimmer rafter
collar beam
common rafter
hip rafter
ridgeboard
header
fly rafter
jack rafter
end rafter
valley rafter
purlin
rafter plate

Collar beam or *collar tie*. A horizontal tie that connects two rafters.

Fascia. The trim board applied to the outside edge of the rafter.

Hip rafters. The rafters that create the third plane of a hip roof.

Jack rafters. The short rafters found on hip roofs or in valleys.

Pitch. The slope of the roof as a ratio relating the rise to the run, generally reported in number of inches per 12 inches. For example, a roof that rises 3 inches for every 12 inches of run would be called a 3:12 pitch.

Pitch line. The elevation where a line drawn from the outside edge of the side wall top plate intersects the center of the rafter–ridge beam intersection.

Purlin. A horizontal framing member that runs at right angles to the rafters or trusses, and, depending on the design, rides on top to act as bracing and a nailer, or on the bottom simply

as structural support.

Rafter. The main framing member.

Ridge beam (also *ridgeboard* or *ridge plate*). The highest member in a roof with two or more slopes (there is no ridgeboard in a roof with a single slope).

Rise. The height of the roof at its peak, minus the height of the side wall, measured at the ridge height.

Run. The distance from the side wall to the ridge (usually half the roof).

Slope. Rise per 12 inches of run, often displayed on plans as a triangle drawn to the side with numbers on its two legs.

Soffit. The outside trim under the rafter overhang.

Span. The distance from side wall to side wall.

Truss. A braced framework of triangles designed to span greater distances than a common rafter.

Flat roofs are always a problem. They all eventually leak, and in heavy-snow areas, they are vulnerable to collapse. At the minimum, a pitch of at least 3 inches of rise for every 12 inches of run (3:12) is desirable. Very steep roofs have their own sets of concerns, and some extra thought needs to go into the design for any pitch that exceeds 6:12. If your heart is set on a steep-roofed design, have an engineer or architect design it for you.

Most roofs overhang the sides of the walls. Overhangs help protect the wall from weather damage, reduce the intensity of summer sun, and/or move snow away from the edge of the building. They can range from a few inches to several feet, depending on what you hope to accomplish with them, but 2 feet is a common amount.

Roofs constructed with a common rafter system require a ridge plate, also known as a ridge beam. Some people install the ridge plate before they begin marking and cutting rafters and then use it to check measurements. Others prefer to use math to prepare all the rafters before raising the ridge plate. The advantage of setting the ridge plate first is that you will have something to work off for checking measurements, but the disadvantage is that if it isn't well secured, a strong wind can wreak havoc. Once the ridge plate is in place — secure it well with bracing to keep it straight — try to work quickly.

Marking and Cutting Rafters

Rafters are a little tricky to cut and prepare. Most are cut in three spots: the two ends of the rafter are cut, and the spot where the rafter rests on the wall is cut with a bird's-mouth cut that allows the rafter to rest flush against the top plate. The bird's-mouth should provide for at least 3 inches of secure contact between the rafter and the top plate of the side wall. Occasionally, a rafter will require one or more additional cuts for complicated rooflines or on shed roofs, which require two bird's-mouth cuts, as the rafter rests on both the front and back walls.

Before marking and cutting rafters, check for the crown on the boards you'll be using. All boards have a crown, or side that is bowed. To find the crown, sight down the length of the board. Then use a pencil and mark the crowned side with an arrow pointing toward the crown. As you mark each rafter, you'll lay the board crown-side up. Sawhorses are handy when marking and cutting rafters, because they need to be laid flat as you work.

Use the straightest piece of wood you have for the first rafter so you can use it as a template. When making the template rafter, mark and cut the top first, because it will snug against the ridge plate on gable roofs and requires a plumb cut. After you cut it, check it to make sure it's correct before cutting the remaining rafters.

Calculating Lengths of Rafters

Before purchasing material for rafters, it helps to run some calculations so that you can get boards of the proper length. When you order lumber for rafters, add at least 1 foot to the calculated length to allow for the cuts you'll need to make, and remember to add length for the overhang(s) if you aren't using flush-cut rafters. In the following examples, we'll assume we want an 18-inch overhang from the wall and that the rafter will snug up against a 1½-inch-thick ridge beam, so we need to account for the thickness of the ridge beam in our calculation by reducing the answer by half the thickness of the ridge beam. There are three basic ways to calculate the length of the rafter, and each has advantages:

1. Hold a board in place and mark it where it needs to be cut. This is easy for small, uncomplicated projects when you are building as you go without the benefit of plans.

2. Use the Pythagorean theorem. (Remember that geometry class they made you take in high school? Well, now you'll get some use out of it.) The theorem says that the square of the length of the hypotenuse of a right triangle equals the sum of the squares of the lengths of the other two sides. Take the square root of the sum of the sides and you will get the length of the hypotenuse. So if your roof will have a run of 12 feet and a rise of 4 feet, the rafter's actual roof length will equal:

$$12^2 + 4^2 = 144 + 16 = 160 \text{ square feet}$$
$$\sqrt{160} = 12.64 \text{ feet}$$

Now calculate the additional length for the overhang:

$$18^2 + 6^2 = 324 + 36 = 360$$
$$\sqrt{360} = 18.97 \text{ inches} \div 12 \text{ inches per foot} = 1.58 \text{ feet}$$

Finally, find the total by adding the two and subtracting for half the width of the ridgeboard (which equals 0.0625 feet when converted from inches):

$$12.64 + 1.58 - 0.06 = 14.16 \text{ feet, or } 14 \text{ feet } 2 \text{ inches}$$

3. Use the rafter table on your large rafter square. This works easily when you're working from plans that provide the slope triangle. Look under the inch mark on the square at the rafter table that relates to your slope at the first line, which reads length of common rafters per foot of run. For example, the plan shows a 4:12 slope. Under the number for 4 inches, you will see *12 65*. Multiply this number, set as a decimal (12.65) by the run in feet and then divide by 12. Let's say the run for this building is 15 feet.

$$12.65 \times 15 = 189.75$$
$$189.75 \div 12 = 15.81 \text{ feet}$$

As above, calculate the length of the additional overhang:

$$12.65 \times 1.5 \text{ (run in feet of the overhang)}$$
$$= 18.98 \text{ inches}$$
$$18.98 \text{ inches} \div 12 \text{ inches per foot}$$
$$= 1.58 \text{ feet}$$

Find the total by adding the two lengths and deducting for half the thickness of the ridge beam:

$$(15.81 + 1.58) - 0.0625 = 17.45 \text{ feet, or}$$
$$17 \text{ feet } 5\frac{3}{8} \text{ inches}$$

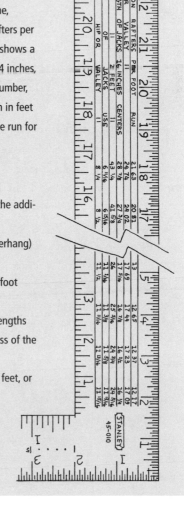

PREPARING A RAFTER

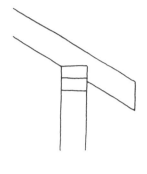

Plumb cut

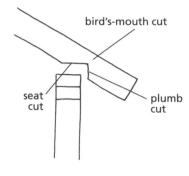

bird's-mouth cut

seat
cut

plumb
cut

Square cut

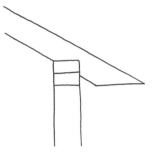

Heel cut

Flush cut

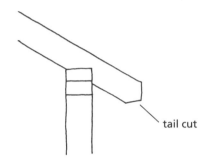

tail cut

Combination cut

Rafter ends are cut in different patterns based on aesthetic considerations. The bird's-mouth cut rests on the top of the outer wall so that at least 3 inches of rafter lies flat against the top of the wall.

Here are the steps for preparing a rafter.

1 Measure the pitch line (see definition on page 202) from the ridge plate if you have temporarily braced it into position, or calculate the measurement using the Pythagorean theorem, but don't forget to adjust for half the width of the ridge plate.

2 Make a plumb cut where the rafter will snug against the ridgeboard. Use the rafter square to get the correct angle for the cut (see box; see illustration A).

3 From the top of the plumb cut, measure a line along the rafter that is the same length as the pitch-line measurement and mark with an arrow. Do the same thing from the bottom of the plumb cut.

4 Draw a line between these two points (it should be a duplicate of the plumb cut, and represents the plumb-cut line for the bird's-mouth).

Making a Plumb Cut with a Rafter Square

To make a plumb cut, use a rafter square. Lay the rafter crown-side up. Place the square so that the unit measure for the rise, as seen on the edge of the tongue (the 16-inch-long side of the square), is on the bottom edge of the board, then swing the square so the unit of measure for the run, as seen on the blade (the 24-inch-long side of the square), is adjacent to the bottom edge. Make sure you consistently use either the marks on the outside edge of the tongue and blade or the marks on the inside edge of the tongue and blade. Mark the line formed by the tongue. Depending on the cut and the width of the board, the point of the square may extend off the board, or it may not make it all the way to the edge, which means

you will have to extend the line to the edge of the board. Let's look at a couple of examples:

1. The triangle on your plans calls for 3:12 roof pitch. Place the 3-inch mark on the tongue on the bottom edge of the board. Swing the square so that the 12-inch mark on the blade is adjacent to the bottom edge of the board. Mark the line created by the tongue.

2. Let's say your plan doesn't provide the triangle but you know your roof will rise 5 feet overall to its peak and have a 15-foot run. Place the 5-inch mark on the tongue against the bottom edge of the board, swing the square so that the 15-inch mark on the blade is against the bottom edge, and mark the line.

5 Keeping the rafter square at the same angle, slide the point along the bottom edge of the rafter toward the plumb cut for the ridge so the 3-inch mark on the blade rests on the plumb-cut line for the bird's-mouth you just drew. Draw a line for the rafter-seat cut; it should be perpendicular to the plumb cut. For steep-roof rafters, 3 inches may be too wide; the cut should be no more than half the width of the rafter (see illustration B).

6 Cut the marked areas. Check to make sure the rafter fits properly, and if it does, use it as a template for quick marking and cutting of the other boards.

The tails can be cut now, but the easiest way is to wait until all the rafters are installed. Snap a chalk line across the tops of the rafters at the point where you will make the cuts, and cut according to the design you have chosen (see the illustrations on page 204).

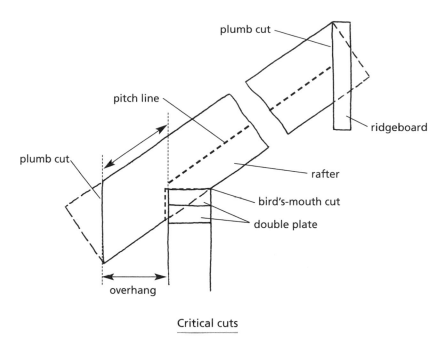

Critical cuts

This illustration shows critical cuts and measurements needed for rafters. The pitch line runs from the center of the rafter's junction with the ridgeboard to the outer corner of the outer wall.

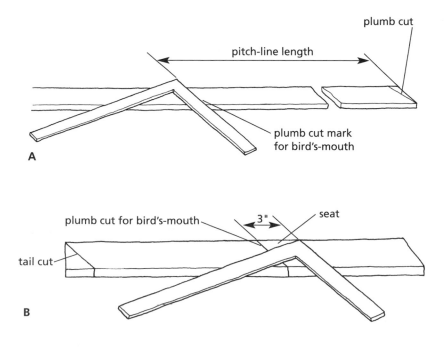

Cut the plumb cut where the rafter butts against the ridgeboard, measure and mark the plumb-cut line for the bird's-mouth cut, measure the seat of the bird's-mouth, and cut the rafter.

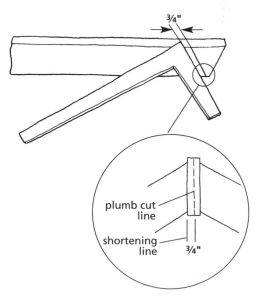

If you don't want to adjust the pitch-line calculations for the width of the ridgeboard, use this trick: Mark the plumb cut and bird's-mouth cut, then move the plumb-cut line closer to the bird's-mouth by half the width of the ridgeboard. For 2-inch nominal lumber, this would be a ¾-inch adjustment.

Installing the Ridge Plate and Rafters

The ridge plate is created from boards that are wider than the rafter boards. For 2x6 rafters, for example, 2x8s are good for the ridge plate. For long runs that require more than one board to create the ridge plate, the boards are spliced together on both sides with pieces of the same size lumber. Some builders prefer to run long ridge plates in a series, marking the first length on the ground, raising it, adding rafters, and then connecting the next stretch that has been marked on the ground. Others make the connections between multiple ridge-plate boards on the ground and raise the entire plate as a single unit.

Before installing the ridge plate, cut it to length and then mark the estimated rafter positions on it while it's still on the ground. If the roof will overhang on the gable ends of the building, make sure you allow for the overhang when cutting the ridge plate. Also mark the rafter positions on the top plates of the side walls.

The ridge plate is initially mounted on vertical bracing posts known as *props*, or *pins*. Two 2x4s nailed together work well; one is cut shorter at the top by an amount equal to the depth of the ridge plate to act as a ledge. Each pin is braced with perpendicular and lateral bracing to get it plumb. Nail the ridge plate securely to the temporary bracing. Use shims as necessary to get the ridge plate level and plumb.

Once the ridge plate is in position, attach the end rafters first. Rafters can be attached with hangers or by nailing. If you choose to nail rafters, attach the top of the first rafter with two 16d nails through the face of the ridge plate. Then attach the rafter on the

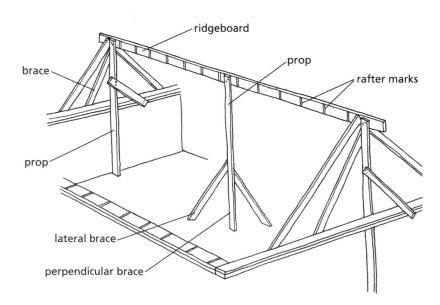

Ridgeboards need to be well braced with temporary bracing. Props and end bracing attached to the outer walls keep the ridgeboard straight while rafters are added.

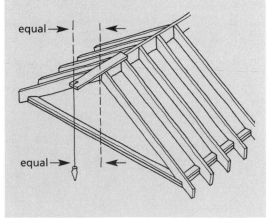

Plumb the Roof

After all the rafters are in place but before the temporary bracing is removed, make a final check.

1. For roofs without a gable-end overhang, tack a 1-foot piece of 1x2 to the ridge plate.
2. To the overhang or the 1x2, hang a plumb bob about 5 inches past the end of the roof. When the plumb bob stops swinging, measure the distance from the top of the rafter and the base. It should be the same. If not, detach one end of the temporary bracing and push or pull the roof until it lines up. Rebrace to hold it plumb until the sheathing is completed.

opposite side of the ridge plate by nailing through the first rafter. As rafter sets are assembled, nail them in at the sidewall. They may be braced with a temporary or permanent collar tie that goes between each set. Once the two end sets are correct, add one or two sets in the middle of the roof to help keep the ridge straight, then begin adding the remaining rafters. Finally, check that everything is still straight and plumb, and add any framing for the building ends under the rafters and the overhang.

Installing Trusses

First, mark all the truss positions on the top plates and nail temporary blocks on the outside walls to keep the end trusses from sliding off the edge of the building as you set them. Place the first end truss upside down between the top plates or supporting girders, with the point facing the floor and about 4 inches from its final position. Use a post (or posts) attached to the top chord with chains or ropes to push the truss upright and into position. Once it's positioned, nail it to the top plates with 16d nails and temporarily brace it with 2x4 props. Work from one end to the other. Once several are in place, you can begin adding lightly tacked purlins as temporary bracing between trusses to help keep them secure.

When all the trusses are in place, check that they are plumb and square. If they are, add the rest and nail securely. For small sheds, purlins may be 1x4s, but for most structures, the purlins should be made of 2x4s.

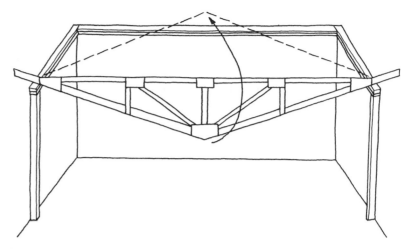

Lay small trusses upside down within the walls of the building, then with the help of a few people, roll them up into position.

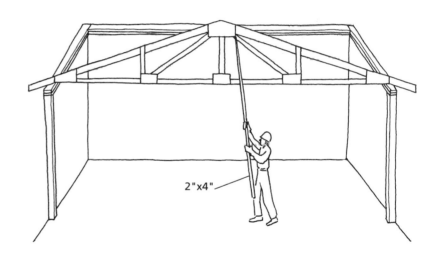

2"x4"

When the truss is in position, nail it in place on the top plate of the outer walls while your helper holds it upright with a 2x4.

SAFETY TIP

For all but the smallest trusses, you will need a crane or tractor with a hoist to get the trusses into position. Erecting large trusses for a big barn requires special skill and so is probably best left to a professional crane operator.

10. FINAL STEPS

Once your building is completely framed, you're ready to move on to the final work — *closing in.* Closing in involves sheathing the walls and roof, installing final siding and roofing, and hanging doors and windows. For animal housing, closing in also includes installing ventilation covers, such as vent caps and cupolas. Windows, doors, and ventilation covers are generally installed over sheathing or framing before finished siding is placed.

For most of us, this is the fun part of the job because the building finally begins to look like the real deal. This is also the part of the project where aesthetics come into play. What kind of siding material do you want? What sort of roofing will you apply? Will you construct your own doors or buy them prehung? What type of color scheme do you have in mind?

As you finish closing in the building, it's time to tackle any finish work your project may require, like electrical, plumbing sys-tems, interior wall cover and trim, and the installation of items to support your livestock operation, such as stall partitions and mangers. This work often takes place over an extended period, as time and funds allow.

CLOSING IN: FINISHING THE ROOF AND SIDEWALLS

Roofs and sidewalls can be finished in any number of ways. But before you get started, to sheathe or not to sheathe is the question. Sheathing is designed to cover the framing; it adds significant strength and structural integrity, and it helps keep drafts and insects out. Generally, half-inch exterior plywood or composites like oriented-strand board are used for sheathing, though if you have your own woodlot, native-lumber boards work well, too. The sheathing is applied horizontally and benefits from the addition of a layer of felt paper or vapor barrier. Offsetting the vertical edges of the sheathing so there isn't a continuous seam helps add structural integrity, thereby reducing shifting and swaying of the building and decreasing the chances that wind will lift the sheathing. Sheathing is nailed with 6d nails, spaced every 6 inches along edges and every 12 inches along studs and rafters.

If you choose not to sheathe, the siding (or metal roofing) will be the final cover on the frame. This option may seem like it will save some money, but it doesn't always do so in the long run. Unsheathed structures require extra sheer panels or bracing to prevent lateral movement. Unsheathed siding and roofing are more vulnerable to failure from weather and the assaults of critters that have a propensity to chew, kick, bash, or peck at new buildings. The long-term result of weather and animal damage is often a sad-looking building with potential structural problems.

One concern in connection with siding and roofing is that it be designed to resist water damage. Water-related problems can be reduced by applying flashing properly, sealing edges, creating an overlapping edge system that causes water to run down, and beveling wood edges so water runs off without working in between boards.

Metal, when used for roofing and siding, offers some advantages when it comes to moisture damage, but even with metal, there are concerns related to water drifting in behind it and rotting the framing. Contraction and expansion of metal panels as the temperature changes causes fasteners (nails and screws) to loosen and back out, which in turn causes the fastener holes to get bigger. Over time, this allows moisture to travel along the panel edge, resulting in leaks that are difficult to detect and repair because they sometimes appear far from their source. To eliminate this problem, apply metal panels with ring-shank nails or screws that incorporate flexible washers. (Many manufacturers offer fasteners designed

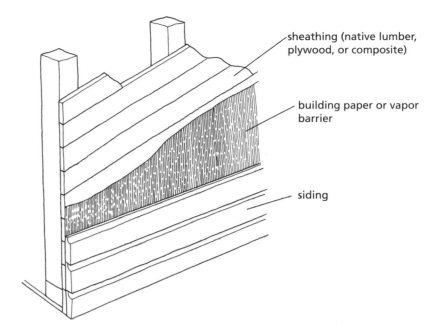

sheathing (native lumber, plywood, or composite)

building paper or vapor barrier

siding

The typical layers of an outer wall: sheathing, which adds strength and structural integrity, building paper or vapor barrier to reduce drafts and control moisture, and final siding.

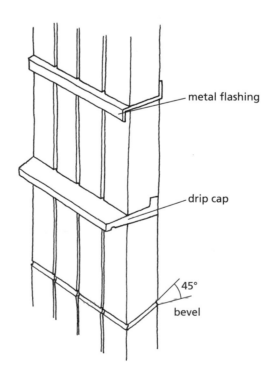

metal flashing

drip cap

45°

bevel

Flashing, drip caps, or beveled edges on wood siding can be used to reduce moisture problems.

TIP

Nailing sheet material (plywood, composite sheathing, or drywall) to framing (studs, headers, joists) can be tricky, because once the sheet is in place, you can't see the framing boards underneath and don't know where to set nails. Avoid this problem by using a chalk line to mark on the sheet material the centers of the studs or other boards you'll be nailing into. This way, you'll know just where to pound those nails.

specifically for their product; the metal panels may also require sealant tape or caulk between overlaps to minimize water seepage between panels.) The washers provide a cushion for panel movement and a water-resistant seam.

Corrugated plastic or fiberglass panels may be used as a substitute in the sidewall or roof system. These translucent panels provide extra light but often require special caulk or treatments to keep them from touching the metal. Follow the manufacturer's recommendations to minimize problems.

Metal panels may rust or discolor more quickly if placed on pressure-treated lumber — particularly if the wood has been treated with pentachlorophenol (also known as penta). Therefore, when using treated lumber, incorporate a layer of builder's felt to protect the metal.

Roofing

Working on roofs can be dangerous. Any time you work far above the ground and on a slope, falls are a real concern. Although you can screw down 2x4 cleats to act as footholds, buying a couple of sets of roof brackets (or roof jacks) is a better idea. They are easy to move, can be used without harming the final roof materials, and they're fairly inexpensive. For safety and convenience, they're well worth the investment.

ROOFING OPTIONS

Type of Roofing	Expected Life	Comments
Metal	3–50 years	Well suited for farm use, tolerates extreme weather, economical
Asphalt shingles	15–40 years	Common, inexpensive, and relatively easy to apply. In extremely windy areas, cheaper styles are likely to fail early; in these applications look for locking, or "T" shingle, designs.
Fiberglass shingles	25–30 years	Similar to asphalt shingles but with longer life expectancy
Rolled roofing	5–15 years	The least-expensive option, but not particularly good-looking. It will last longest if the edges are recoated every couple of years with roof cement. It will also last longer and protect better if two layers are used with offset seams.
Wooden shingles	50+ years	Wooden shingles and shakes are good-looking but initially expensive and slow to apply. In the past, their vulnerability to fire was a major disadvantage, but new ones are pressure-treated with a retardant.
Slate and tile	100+ years	The most expensive up front, slate and tile roofs will last and last. They can withstand almost anything Mother Nature throws at them. Their weight adds significantly to the dead load.
Rubber roofing	25–35 years	This is a specialty product that a contractor must install, but it's the best option on flat roofs.

Roof jacks are an excellent investment for doing a roofing job. They are inexpensive and provide a good working platform.

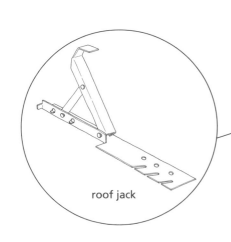

roof jack

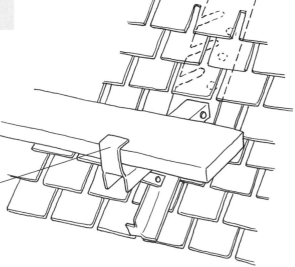

Materials

How you finish the roof depends somewhat on its slope. A steep roof (greater than 4:12) can be covered with any kind of roofing material. For roofs with 2:12 to 4:12 slopes, don't use wood shingles, slate, or tiles, as these are more likely to sustain water damage when used in this application. For roofs with slopes less than 2:12, use rolled roofing and roof cement to seal the seams, or rubber roof coating (which has to be done by a contractor).

Flashing is applied before placing the final roofing material. Flashing is thin metal or plastic that's applied at all seams, including around pipes, vents, chimneys, or a cupola. It's designed to keep moisture from seeping down to the wood underneath the final roof cover. Metal flashing must be attached with fasteners that are compatible. For example, if aluminum flashing were fastened with steel nails, corrosion would occur where the steel and aluminum touch. Check with the supplier for recommended fasteners, or use fasteners made of the same metal as the flashing.

Metal roofing is probably the most common type of roof finish for barns. It's fireproof and allows snow to slide off quickly. For uninsulated structures, the metal panels are generally placed on purlins, or horizontal roof

FLASHING

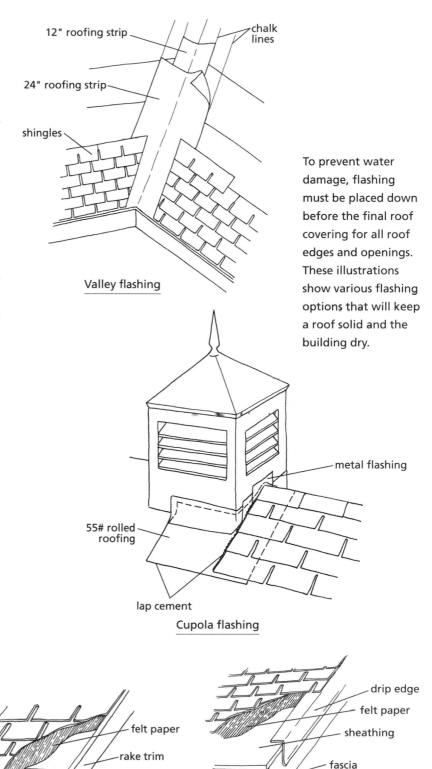

To prevent water damage, flashing must be placed down before the final roof covering for all roof edges and openings. These illustrations show various flashing options that will keep a roof solid and the building dry.

12" roofing strip

chalk lines

24" roofing strip

shingles

Valley flashing

metal flashing

55# rolled roofing

lap cement

Cupola flashing

cut shingle

pipe sleeve

stack flange

Vent collar

felt paper

rake trim

rake flashing

Rake

drip edge

felt paper

sheathing

fascia

rake trim

Drip edge

METAL PANELS

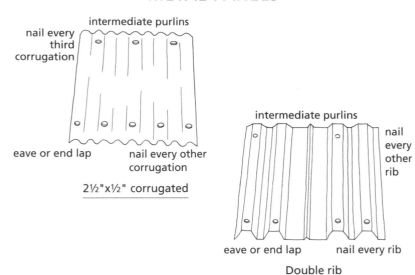

nail every third corrugation
intermediate purlins
eave or end lap
nail every other corrugation

2½"x½" corrugated

intermediate purlins
nail every other rib
eave or end lap
nail every rib

Double rib

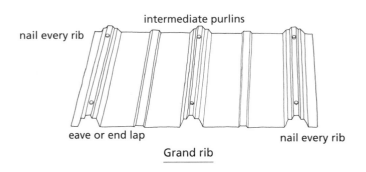

nail every rib
intermediate purlins
eave or end lap
nail every rib

Grand rib

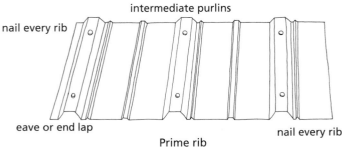

nail every rib
intermediate purlins
eave or end lap
nail every rib

Prime rib

There are many kinds of metal panels; each type has specific requirements for fasteners and fastening layouts.

Fasteners typically go through the "rib" on metal panels. They should snug flush against the rib without indenting it.

members, that run in continuous strips across the roof at 2-foot intervals, rather than on sheathing, though it works perfectly well over sheathing for insulated structures. The least-expensive type of metal roofing is farmer's tin, or corrugated galvanized sheets, but today you can also choose from a number of metal panel products available in a wide variety of factory-applied colors and patterns (some are designed to look like tile, shake, or shingle roofing). They are good-looking, long lasting, and reasonably priced.

Laying Out Roof Panels

The sequence for panel layout is as follows: Start at the bottom corner on the downwind side so that vertical laps are away from the prevailing wind direction. Add the next horizontal layer, maintaining a 6-inch overlap (or an overlap specified by the manufacturer) of the first sheet. Continue moving up to the peak, then move over and repeat. Allow at least a 2-inch overhang on the eaves to provide a drip edge. When the sheets have all been applied, add ridge caps and pipe boots.

Applying Sheathing and Roof Felt

Roofs on insulated structures, or those that are going to be covered with rolled roofing, shingles, or tiles, require sheathing. Exterior

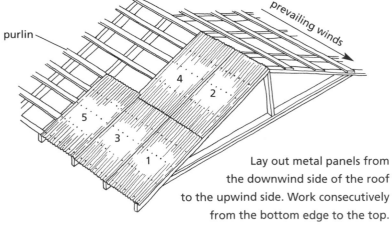

purlin

prevailing winds

Lay out metal panels from the downwind side of the roof to the upwind side. Work consecutively from the bottom edge to the top.

plywood (½ inch) is the most common choice, but 1-inch boards also work well.

Apply sheets of plywood in a staggered pattern so the seams don't join up. It's easiest to start at the bottom of the roof and work all the way across on the first horizontal layer, then move up the roof. Plywood should be nailed down every 12 inches along the rafters (6 inches on the edges) using ring-shank nails. As you work up, use roof brackets or cleats to provide safe footing.

As soon as possible after the sheathing is up, apply 30-pound rolled roof felt and staple securely. Roof felt helps to protect the roof from the ravages of weather. The rolls come in 36-inch-wide sheets; each sheet should overlap by at least 3 inches or as specified by the manufacturer.

Applying Rolled Roofing

Rolled roofing is an asphalt-impregnated product. Like roof felt, it comes in 36-inch rolls. It is inexpensive and quick to apply, but you get what you pay for: it's the shortest lived of any roofing finish and its appearance is plain Jane. On roofs with slopes less than 1:12, however, several layers of rolled roofing "painted" with an asphalt roof coating can effectively keep out moisture. (On low-slope roofs, the roof coating will need to be reapplied every couple of years to maintain water resistance.)

The best time to apply rolled roofing is during the heat of summer. It cracks and breaks when temperatures fall below 45°F and is stiff and hard to work with if temperatures are below 65°F. Roof cement, which is also easiest to apply when the ambient temperature is high, is added between seams and between layers. It goes on the roof quickly, but it's messy work.

Installing Composition Shingles

Composition shingles (asphalt or fiberglass) are easy to install and come in different weights. For longevity, select heavier shingles; they withstand sun, wind, and temperature changes better than lighter shingles. For most locations, standard shingles are adequate; they come in a strip and may be solid edged or designed with keys, or tabs, that are separated by a notch and yield a pattern when laid. If you are building in an area that's subject to strong winds, consider using interlocking shingles, which have specially designed keys that weave together to resist strong winds.

Set a starter course along the lower edge of the roof with the keys pointing up the roof.

TIP

For cutting a few metal panels, handheld, straight-edged tin snips will work. If you need to cut many, rent or purchase electric shears or nibblers. Cut plain galvanized panels with a circular saw equipped with a metal cutting blade, but don't use this approach on panels that are enameled by the manufacturer (colored sheets); doing so will kick up hot metal filings that embed in the painted surface and welcome rust.

TIP

Before applying sheathing and roof felt, snap a chalk line 47½ inches up from the end of the rafter, or if there will be a fascia board 48 inches from the center of the fascia board. This gives you a starting point for the first row of sheathing and will help keep your work straight.

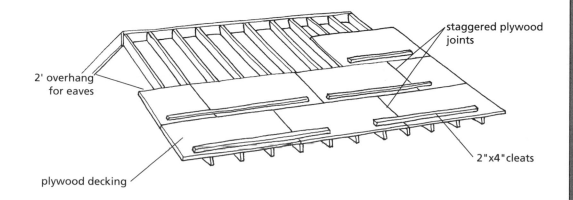

Sheathing should be laid out in a staggered pattern so the seams don't match up. For safety, nail down 2x4s as temporary cleats; use double-headed nails for easy removal.

COMPOSITION SHINGLES

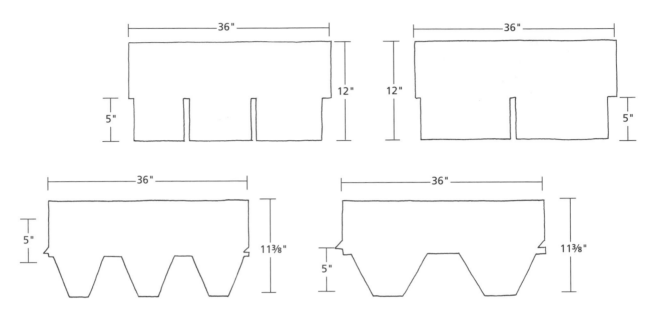

36"

12"

5"

36"

12"

5"

36"

5"

11⅜"

36"

5"

11⅜"

Standard shingles

Composition shingles come in many styles. Standard shingles are the least expensive. Interlocking shingles are more expensive but resist wind damage far better than do standard shingles.

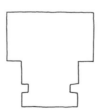

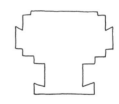

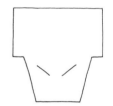

Interlocking shingles

INSTALLING SHINGLES

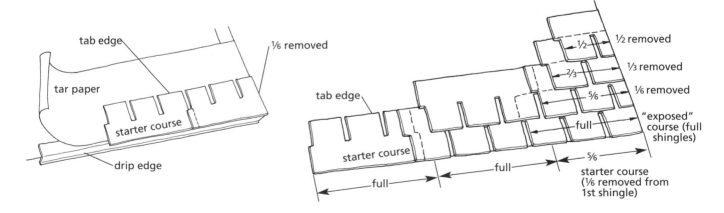

tab edge

⅙ removed

tar paper

starter course

drip edge

tab edge

½ removed

½

⅓ removed

⅔

⅙ removed

⅚

full

"exposed" course (full shingles)

full

⅚

starter course

full

full

starter course (⅙ removed from 1st shingle)

Apply tar paper and flashing. Lay a starter course of shingles, upside down, along the bottom edge of the roof.

Place the first full row of shingles on top of the starter course. Then, to get the pattern correct, cut edge shingles in subsequent rows. The cutting pattern for a three-tab standard shingle is shown in this illustration.

Then place the first row of shingles that will show over this course, with the keys in their proper position, pointing down. The starter course provides a slight drip edge.

Continue, building layer upon layer. Depending on the style of shingle you use, you may need to trim the edges of the first shingle in each horizontal layer so that you have an overlapping pattern. The most common trimming pattern is a six-layer trim, which results in the notches being spaced equally between the tabs above and below. Once the shingles have all been applied, run shingles over the ridge in an on-edge pattern, laid in the direction of the prevailing wind, to finish the ridge.

Installing Wooden Shingles

Cut (or split) from cedar, redwood, cypress, or western red cedar, wooden shingles (or shakes) will shed water for years and add superior insulation values. Wood shingles are not naturally fire resistant (some are treated with a retardant); they may not meet fire codes depending on the region of the country where you live. Shingles are set about ¼ inch apart (a carpenter's pencil works well to create this gap). Traditionally, each layer of shingles received its own layer of roofer's felt, but today a single layer of felt can be applied to the entire roof; then plastic mesh is laid over the felt before the shingles are applied. These two steps ensure much-needed air circulation that allows the wood to dry after rains and snow and also allows for the slight expansion and contraction that takes place as temperatures change.

Siding Options

Your choice of siding will depend on your budget and the aesthetic requirements of your neighborhood. If you live in a truly rural area, farmer's tin or board-and-batten will be fine, but if you live in a suburban/urban edge area and have neighbors, you may need to build a

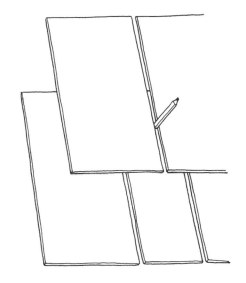

Wooden shingles are laid over individual layers of roof felt or over plastic mesh that covers the whole roof. Use the flat side of a carpenter's pencil as a spacing guide when laying wooden shingles.

custom-metal building or use horizontal board siding that is more in keeping with the style of area houses.

Metal siding, vertical board siding, or horizontal board siding can be added on top of sheathing. When sheathing will be covered by another siding, ½-inch plywood or oriented strandboard is heavy enough. Adding a layer of felt paper or vapor barrier between the sheathing and siding will probably be required by code in the eastern United States but not in the arid West.

Exterior Plywood Sheathing

Exterior plywood (¾-inch) sheathing can, when painted for sidewalls, be used as the final siding. Exterior plywood is also available in textured panels designed to look like rough-sawn planks, batten-and-board, and so on. Another approach to improve the finished look of plywood sheathing is to nail 1x2 boards as battens. Rough-sawn plywood is popular for this option.

Metal Siding

Metal siding without sheathing is popular for barns; it is economical and requires mini-

SHAKE OR SHINGLE?

What's the difference between wooden shingles and wooden shakes? Shingles are sawed; shakes are split.

When metal siding is used without sheathing, include a stall lining on the inside of the structure to protect the metal, and your animals, from injury. Stall linings can be made from ¾-inch plywood or 1-inch lumber and should be 4 feet tall.

mum maintenance, but it is particularly susceptible to the fate of unsightliness, and animals (especially horses) can occasionally kick all the way through it, resulting in serious injury. To keep unsheathed metal looking better and to reduce the chances of animal injuries, build a 4-foot-tall stall liner of ¾-inch plywood or 1-inch lumber (6-inch boards work well) on the inside of the wall of any areas to which animals will have regular access.

Vertical Siding

Vertical siding, like board-and-batten siding, is appropriate for barns — and performs best when used as unsheathed siding. Board-and-batten is made with 1x6 or 1x8 boards and 1x2s or 1x3s as the battens. An alternative is board on board, where each board overlaps its neighbor by about an inch. Vertical siding requires additional nailers set horizontally. These can be blocked 2x4s between the studs or (the less expensive and easier approach) 1x3 straps nailed to the outside of the frame.

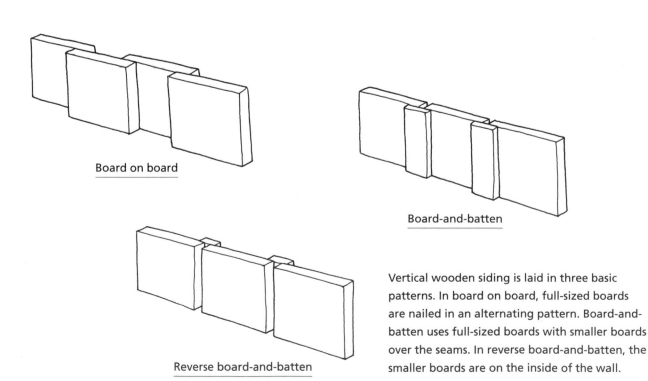

Board on board

Board-and-batten

Reverse board-and-batten

Vertical wooden siding is laid in three basic patterns. In board on board, full-sized boards are nailed in an alternating pattern. Board-and-batten uses full-sized boards with smaller boards over the seams. In reverse board-and-batten, the smaller boards are on the inside of the wall.

Horizontal Siding

Horizontal siding is the most common type for houses, so for those who want to match the look of their home, it's probably the siding of choice. There are a variety of styles for horizontal siding. All horizontal siding starts at the base of the wall and works up toward the top so that each horizontal board overlaps the one below it. This helps reduce water damage.

Native Lumber

Locally sawn, native lumber is perfectly suitable for barn siding. It does not cost as much as finished lumber, and it gives the barn a kind of rustic charm and character. Although the wood doesn't need to be fully dried to be used as siding, it will benefit from a couple of months of air-drying.

Fiber-Cement Planking

Another product making its way into the construction universe and that shows real promise for siding is fiber-cement planking. It is composed of cement, sand, and cellulose fiber that have been autoclaved (cured with pressurized steam). By adding fiber to cement, its strength and dimensional stability are increased, and its tendency to crack, an inherent problem with regular concrete, is decreased. The planks come in widths ranging from 5¼ inches to 12 inches and are about 5/16 inch thick. They are handled much the same as regular boards, can be cut with a carbide-tipped saw blade, and they take paint similar to the way regular wood does. Manufacturers of fiber-cement planking emphasize that their product is resistant to rot and fungus, as well as termite infestations, and that their product has excellent weathering characteristics, strength, and impact resistance. The installed costs of fiber cement are reported to be less than for traditional masonry or synthetic stucco, equal to or less than those of hardboard siding, and more than those of vinyl siding.

Vinyl Siding

Vinyl siding is popular for home construction, but it tends to be a poor choice for barns. Vinyl is subject to chewing (or pecking) by animals, and once a hole gets started on a corner, it grows quickly.

Hanging Doors and Windows

Factory prehung doors and windows can be installed quickly, used doors and windows can be adapted, or you can build your own. Prehung doors and windows are set into rough openings that are slightly larger (usually ½ inch larger all around) than the jambs provided by the manufacturer. Use shims to set the door or window square in the roughed-in opening before nailing it into place with 12d finish nails. There are several steps you can take to reduce drafts around prehung doors or windows:

- Include a layer of builder's felt around the edges.
- Squirt a bead of caulk all around the edge where the casing or nailing flange and wall or frame meet.
- Fill gaps between the roughed-in opening and the jambs with insulation or caulk.

Prehung doors work well for man doors (doors for people), but they are not readily available in sizes wide enough for large animals like horses and cattle to pass through. If and when you do find them in larger sizes, they are very costly. A prehung door comes braced so that it remains square. Don't remove the bracing until you absolutely have to; the door can quickly shift out of square, and it's difficult to get it back into position. Consider purchasing exterior-type doors even when you need interior doors. Most modern interior doors have a hollow core with a thin veneer over them; they may be suitable for household use, but they can't handle the hard wear and tear to which an interior barn door is subjected.

TYPES OF DOORS

Basic doors include hinged double doors, sliding doors, and overhead doors for large openings that accommodate equipment, and hinged single or Dutch doors for openings that animals or people use.

Hinged double door

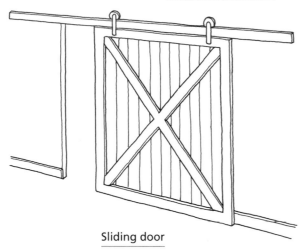

Sliding door

Homemade doors can be constructed of plywood or 1x4 lumber, and there are several bracing designs you can use. For unheated barns, surface-mounted sliding doors work fine and eliminate the need to construct a finish jamb; for buildings that require a better seal, you can build your own jamb. Large doors that permit access for vehicles or tractors are usually hung on a track system so they can be slid out of the way, though overhead doors (with or without automatic openers) are an option when budget isn't a concern. Hinged doors for these large openings are usually heavy and quickly begin to sag.

Windows should be high enough on the wall that animals can't get at them or covered with a protective metal grill. Wooden-framed windows are nailed into the rough opening through the casing along their outer edge; metal, vinyl, and vinyl-clad wooden windows generally have a perforated flange for nailing and, for the flange to sit properly, may require the sheathing to be trimmed back from the opening. These flanges may be brittle, particularly in cold weather.

One of the best window choices for barns is a hopper-style window placed high on the

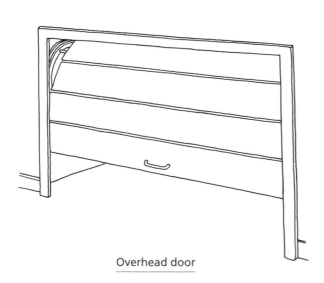

Overhead door

Hinged single door

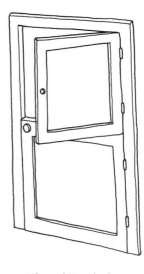

Hinged Dutch door

Installing a Window

Although it's possible to install windows (or prehung doors) solo, it's a whole lot easier if you have a helper nearby to adjust the shims. The following steps explain how to install a new wooden window, although setting a prehung door involves essentially the same steps.

1 Mark the sheathing for the window opening to be cut.

2 Set the depth of the circular-saw blade slightly greater than the depth of the sheathing and plunge the blade into the board to begin the cut.

3 Set a drill with an expansion bit inside the first corner and cut a hole large enough to accommodate the blade of a keyhole saw. Work the line with the keyhole saw until it is long enough to set in the circular-saw blade, then finish the cut with the circular saw.

4 Wrap an 8-inch strip of felt around the inside edge of the window frame and staple. Apply caulk.

5 Set the window into the opening and, with a single nail, tack it temporarily in the corner of the upper casing. In order to keep the window square, leave bracing or reinforcing blocks in place on the window until the job is completed.

6 Have your helper use tapered shims to adjust the window height so it's level. A torpedo level works well when leveling windows.

7 Once the window is level, secure it in position by tacking nails in each corner.

8 Recheck the window's position to ensure it is plumb and level. Check squareness by measuring diagonals. Operate the window to make sure it opens and closes easily.

9 Working on the inside, combine pairs of shims to form flat shims and insert them in the gaps between the jambs and the framing at 12-inch intervals and near each corner, so they're snug. Use a straight edge (the long side of your roofer's square will work well for this) to make sure the jambs aren't bowed. Adjust shims, if necessary, until the jambs are flat.

10 At each shim set, drill a pilot hole in the jamb, then nail an 8d casing nail through the jamb and shims into the rough-opening framing. Using a nail set will allow you to secure the nails flush with the surface of the wood without denting it.

11 Remove the tacked corner nails. Apply a dot of caulk on each hole to seal it.

12 Stuff insulation into the gap between the window and the rough jamb.

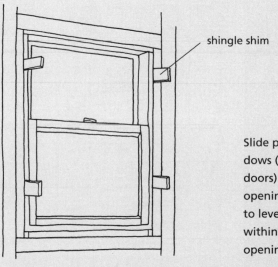

shingle shim

Slide prehung windows (or prehung doors) into a rough opening and use shims to level and straighten within the rough opening.

wall. Hinged at the bottom, these windows provide good ventilation while avoiding a draft directly on animals. Combined with a "ventilation flue," they are low-cost air fresheners.

Recycled windows work well for barns. We have purchased decent windows at garage sales, at auctions, and from building recyclers. An old single-sash window can be quickly adapted for use as an inverted hopper. It also can be adapted to operate in an easily constructed slide.

Installing Vents and Ventilation Covers
Ventilation covers should be installed before final siding. Often they are little more than plastic or metal screens and louvers in gables,

TYPES OF BRACING

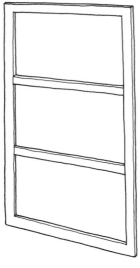

Brace and frame

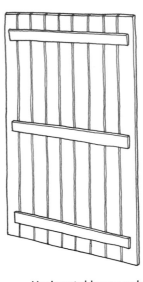

Horizontal brace only

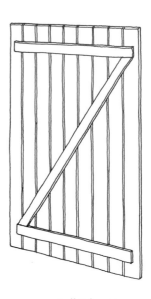

Full Z-brace

Bracing is used to stabilize homemade doors. The larger the door, the more bracing it will require.

Double Z-brace

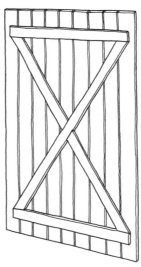

Full X-brace

soffits, walls, and the roof, but for barns where livestock will be housed full time, consider including commercial ridge vents on the roof and exhaust fans in the walls. Because of natural airflows, open doors and windows can't remove the moist air and ammonia fumes on a hot, still day if there isn't a place for them to exit at the top of the roof. Roof vents, including cupolas, are cut into the roof much the same way as windows are cut into a sidewall. To reduce insect pests, staple screen behind all vent covers and on the inside of any cupola you include.

As a rule of thumb, provide approximately 3 inches of screen-covered eave vent for every 10 feet of building width, along the entire length of the eave, and ridge vents or cupolas at the rate of approximately 1 square foot of usable opening per 120 square feet of floor space. The eave vent is maintained in an open position year-round, but it can have a hinged flap or slide-type cover that fits into place to close off all but a narrow strip for extremely bad weather, such as periods of heavy, driving snow. When using cupolas instead of ridge vents on barns longer than 60 feet, opt for multiple units.

For example, on a 36-foot-long barn that's 24 feet wide, you need at least 7.2 inches (2.4 [the number of 10-foot widths to the building] x 3 [inches of screen-covered eave vent required per 10-foot width]) of eave vent running the length of the building, so you should use a 3.5- to 4-inch strip on each side of the building for the entire length of the building. For this same building, there is 864 square feet of floor space, so allow 7.2 square feet (864 ÷ 120) of usable opening for roof vents or cupolas. One 33-inch square cupola provides 7.5 square feet of roof opening, but make sure the louvered area also provides at least 7.2 square feet of opening. If instead the barn was 60 feet by 36 feet, you would need at least 5.4 inches of eave vent down both sides (3.6 [10-foot widths] x 3 ÷ 2), and roof vents or cupolas

TYPES OF WINDOWS

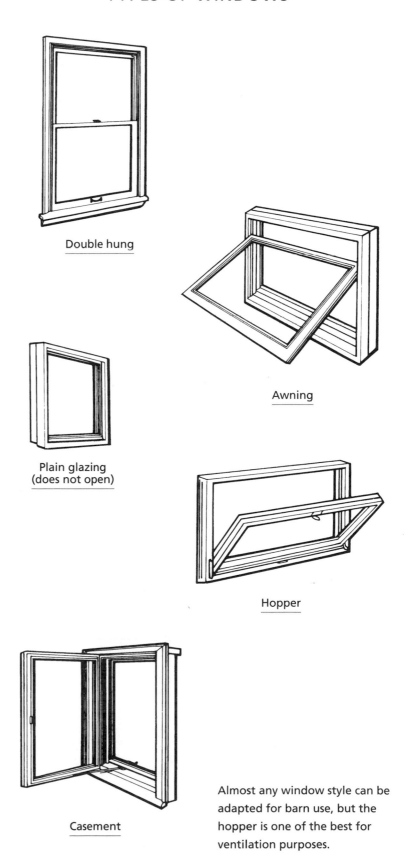

Double hung

Awning

Plain glazing
(does not open)

Hopper

Casement

Almost any window style can be adapted for barn use, but the hopper is one of the best for ventilation purposes.

Placing hopper-style windows high on the wall *(right)* and using ventilation chimneys *(left)* helps keep barn air fresh without causing animal-killing drafts.

sized to provide 18 square feet of opening (60 x 36 ÷ 120). A 6-inch-wide roof vent along the entire length of the roof would do the trick, as would two 25-inch square cupolas.

Door, window, and side-wall openings that can be opened in warm weather and closed in cold weather should be available to provide the equivalent of at least 20 percent of the floor area on the long walls. In the 36-foot-long by 24-foot-wide barn from the previous example, the total space available as closable wall openings should be at least 172 square feet (36 x 24 x 0.2). On gable, gambrel, or monitor roof barns, design the openings so they are of comparable length on each of the long walls; for shed-type buildings, create approximately one-third of the long-wall openings on the back wall and two-thirds on the front wall.

tight roof deck

air in

eave vent door (optional)

wall vent door (optional)

insulation optional in wall, vent door, and ceiling

exposed 6"x6" pressure-treated post

siding

2" stall lining

Wall and eave vent

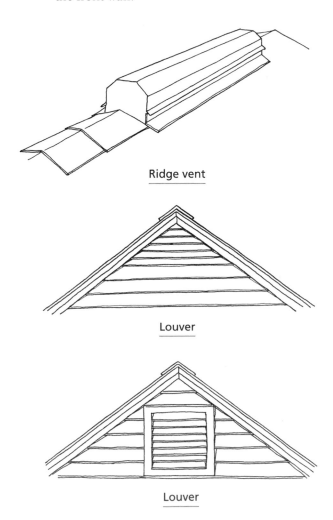

Ridge vent

Louver

Louver

Louvers, ridge vents, eave vents, and wall vents are all good natural ventilation devices. In very cold climates, incorporate optional vent covers that can be closed when temperatures plummet below zero.

Dan Herman

Dan Herman is a contractor in Colorado with more than 31 years in the business. He has built virtually every kind of building, including about twenty barns and a $15 million commercial equestrian center that features several large stables, an indoor arena, and other facilities.

Early in his career, Dan learned an approach to hanging windows and doors that was used by framers in Southern California; it didn't rely on shims and could be readily done by one person. In lieu of shims, Dan recommends "floating trimmers." What follows is his approach.

1 Create a rough opening for the window or door, with the king studs placed so that the sides of the opening are equivalent to the size of the unit to be installed, plus the size of floating trimmers (which are made from the same size lumber as the studs), plus 1 inch. For example, if the frame of the unit is 30 inches and you are using 2x4 studs, the opening between king studs will be 34 inches (30 inches for the unit, 3 inches for two 2x4 floating trimmers, and 1 inch for the floating area). For very large doors or windows, add an extra trimmer above the sill and nail it to the jack studs on each side to help support the header. With the extra trimmers, the rough opening in this example would be 37 inches, or 7 inches larger than the unit you're installing.

> *". . . you tend not to have issues with moving parts"*

2 Install the headers (for windows or doors) and the sill plates (for windows) with cripple studs. (Dan uses two sill plates for a stronger sill.)

3 Cut and set the floating trimmers within the opening so there is about ⅛ inch extra space around the window or door. Make sure each floating trimmer is square and plumb, then secure with "locking nails." (Dan typically uses two locking nails on each floating trimmer for small/medium windows and three or more for doors or large windows.)

4 Gently slide the unit into the opening and nail it in place with finish nails.

According to Dan, "A floating trimmer creates solid backing for the window or door, and it won't come loose or move around as shims do. Because the floating trimmers don't move or shift, you tend not to have issues with moving parts, like slides, hinges, and latches. With shims, you only have small places to nail to, but with floating trimmers you have good material to nail external trim to and on which to place a rounded drywall corner."

Install insulation in the small gap between the studs and the floating trimmer.

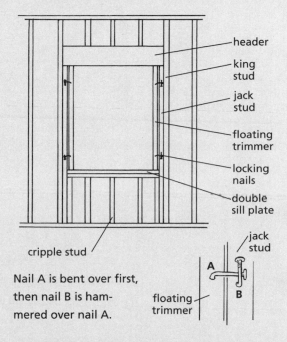

header
king stud
jack stud
floating trimmer
locking nails
double sill plate
cripple stud
jack stud
floating trimmer
A
B

Nail A is bent over first, then nail B is hammered over nail A.

Building a Cupola

A cupola and weather vane are quintessential barn adornments in the minds of most people, but they actually serve practical purposes in ventilation. Several companies sell prebuilt cupolas, and some lumberyards carry kits for them, but you can also build one yourself.

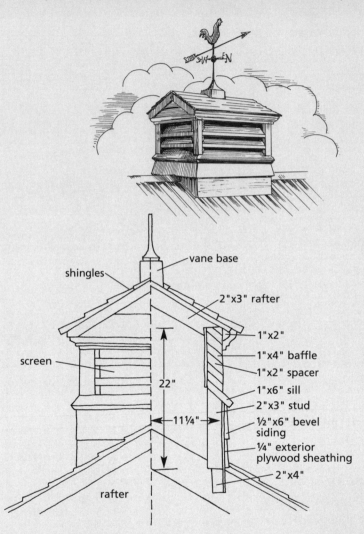

shingles

vane base

2"x3" rafter

screen

1"x2"

1"x4" baffle

1"x2" spacer

22"

1"x6" sill

2"x3" stud

11¼"

½"x6" bevel siding

¼" exterior plywood sheathing

2"x4"

rafter

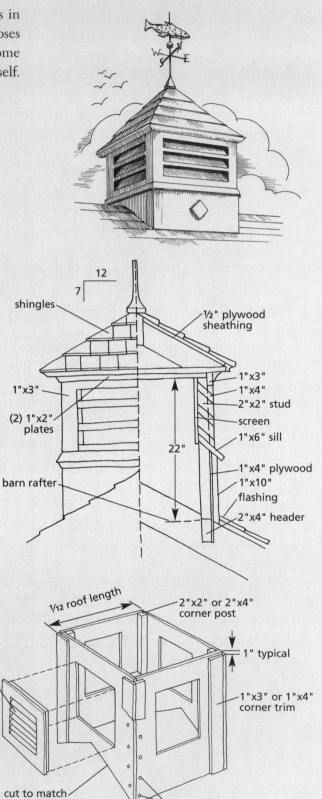

12

7

shingles

½" plywood sheathing

1"x3"

(2) 1"x2" plates

1"x3"

1"x4"

2"x2" stud

screen

1"x6" sill

22"

barn rafter

1"x4" plywood

1"x10"

flashing

2"x4" header

1/12 roof length

2"x2" or 2"x4" corner post

1" typical

manufactured vent louvered and screened

1"x3" or 1"x4" corner trim

cut to match roof pitch

screws 4" on center at corners

Each plan corresponds to the cupola design above it. These are somewhat complicated to build, but the finished product is worth the extra effort.

A quicker, easier approach to constructing cupolas is to create the basic structure with plywood and manufactured louvers.

EXTERIOR FINISH WORK

Now you're ready to tackle finish work, like adding trim, installing gutters or hardware on doors and windows, and painting, but these jobs can be undertaken at your own pace. If you have to do electrical and plumbing work or interior finishing like drywall for office or bathroom walls, you may want to hold off on the exterior finish work, but if your building requires painting, don't wait too long before applying at least one coat of primer.

I use the word *painting* loosely, as you have a choice of paint or stain. Paint is a suspension of pigments in latex or oil. It forms a film over the surface of the wood (or whatever surface it is applied to) that resists sunlight and water. Acrylic latex-based paints are preferred for exterior use because the acrylic keeps the paint flexible and durable, resulting in less cracking and peeling, with better lifetime performance.

Stain comes in semitransparent varieties (which can include pigment) that allow the grain of the wood to show through, or solid, colored types for a paintlike finish. Semitransparent stains have the advantage of penetrating the wood deeply, which seems to offer better protection, and resisting UV radiation better than other wood treatments. They also work best on native or other rough-surfaced wood. They don't bubble, crack, or peel the way paint does as it ages, so recoating later is an easier job.

Whichever coating you choose, buy a high-quality brand that specifies exterior use, and make sure the surface is dry before applying it. Paint and solid stain are best applied midday during hot weather, but semitransparent stain works better when applied during slightly cooler times, such as in spring and fall.

Paint Safety

Almost all coatings have some hazard associated with them. Oil- and alkyd-based paints are potentially the most hazardous to humans and the environment, and they are combustible.

- Allow time for off-gassing of fumes, even if you use the less hazardous paints and finishes listed below. For indoor painting, it's best to paint when it's warm enough to keep windows open, with fans turned on to push the fumes out and bring fresh air in. Whatever process is used, proper ventilation is important. Paint fumes can be hazardous if inhaled and can also be an explosion hazard.

- Pregnant women should not paint! Fumes inhaled by the mother can pass through the placenta and harm a developing baby.

Surface	Latex paints	Oil-based paints	Epoxy-based paints	Latex stains	Oil-based stains
Unfinished wood	✓+	✓+	X	✓	✓
Painted wood	✓	✓	✓	X	X
Hardboard siding	✓	✓	✓	X	X
Stained wood	X	X	X	✓	✓
Redwood siding	X	X	X	✓	✓
Cedar siding	✓+	✓+	X	✓	✓
Metal siding	✓	✓	X	X	X
Concrete	✓+	✓+	✓+	X	X

✓ = suitable for use in this application

✓+ = suitable for use but should be painted with primer first

X = not suitable in this application

• Wear nitrile gloves, a long-sleeved shirt, and long pants when you paint, and *always* wear a mask that is specifically designated for painting. Don't use a dust mask — it won't keep out fumes and, in fact, can capture them near the mouth and nose.

• When shopping for paints, look for the following types of paints, which emit fewer fumes than conventional — especially oil-based — paints.

—Low-VOC (volatile organic compound) paints, stains, and sealants are formulated so they meet California guidelines, which, in terms of allowable levels of carcinogenic chemicals, are stricter than federal regulations. While these paints may still emit fumes, they do so usually at far lower levels than conventional oil-based and water-based paints. Some companies even formulate their coatings so that they won't emit any fumes.

—"VOC-free" or "no VOC" paints don't emit any fumes but generally cost more.

—"Low-biocide" paints are 90 to 95 percent free of biocides, which include dangerous preservatives and fungicides. However, low-biocide paints are more prone to mold and should not be used in damp areas like basements or bathrooms, unless there's plenty of sunlight and ventilation.

—Natural or organic paints are made from citrus and other plant oils. They are free of synthetic chemicals. Water-based natural paints are, therefore, less resistant to mildews and molds. Natural oil-based enamel paints can be used in humid areas instead. These paints can contain aromatic ingredients such as the citrus-based solvent d-limonene, turpentine, tung oil, and pine resins, which can cause reactions in sensitive people.

• Avoid throwing old paint in the trash unless it has completely dried. Liquid paint should be dropped off at a hazardous waste drop-off site. To find a drop-off site in your community, search by zip code at Earth's 911 (see resources). Adhere to the following steps to make disposal safer for the environment.

—For small quantities: Either brush all the remaining paint onto a piece of cardboard or newspaper and allow the empty, unlidded can to dry, or leave the paint in the uncovered can to dry. Drying should be done in a well-ventilated area protected from open flames, children, pets, and rain. If the amount of paint occupies less than one-fourth the container, the paint should be able to dry in the can. Stirring the paint every few days will speed up the process.

—For large quantities: Pour half-inch layers of paint into a plastic-lined cardboard box. The paint should be allowed to dry before adding the next layer. Mixing in an absorbent material like kitty litter or sawdust will speed the drying process. Or donate partial cans to art programs, theater groups, or to a paint "bulking program" that mixes old paints and redistributes them.

—For paint that has separated: Pour the clear liquid off the top into a plastic-lined cardboard box. Add enough kitty litter or other absorbent material to take up all the liquid. Dry the remaining paint by following the instructions above.

—For paint-saturated rags: Place in kitty litter and dispose of in trash.

—Once dried, paint may be included with the trash. Whenever setting a dried paint can out for trash collection, leave the lid off so the collector can see that the paint has hardened.

—For paint thinners: Used thinners, turpentine, and mineral spirits can be reused. Pour into a clear glass container and seal with a tight lid. Allow the paint particles to settle to the bottom. Pour the clear liquid into the original container for reuse, then dispose of the residue after allowing it to dry. Adding an absorbent material such as kitty litter or sawdust will speed up the drying.

ELECTRICAL SYSTEMS

Although small buildings might not require electrical service, most full-sized barns will require electricity — at least to power lights and small appliances, such as a radio or power tool. For a minimal application, you may be able to run a line from your existing main breaker panel (see below for more information) to a new subpanel with its own disconnect switch; however, for larger services, you should install a new main breaker panel or service. If you will have a number of outbuildings serviced by electricity, consider installing a yard pole with the main service entrance and meter mounted on it, and then branch off to each building's main breaker panel with feeder lines.

The service lines that drop from the power grid to the meter are installed by the local utility company, but from the meter on, lines are the owner's responsibility. Almost every state or municipality requires an electrical permit and inspection, even if they don't require other building permits, and some require that all work be done by a licensed electrician or that a homeowner who wants to do his or her own work pass a test to demonstrate competency before undertaking any electrical work. There's good reason for their concern: many fires and deaths (including animals) have been caused by electrical system failures. For example, in one Nebraska study, more than 70 percent of on-farm fires were the result of electrical malfunctions, and 100 percent of the farms surveyed in the study had at least one condition related to their electrical system that was considered a "serious threat" to safety.

In spite of these issues, most barn wiring doesn't have to be onerous, and you can do it yourself, though if you have not worked with electricity before, I strongly recommend having a licensed electrician make the connection from the meter to the main breaker panel. My best advice for those who are going to do it themselves follows.

- Take your time. Rushed electrical work is dangerous electrical work.

- Use only equipment and devices listed by Underwriters Laboratories (UL) and/or the National Electrical Manufacturers Association (NEMA). These organizations list items that have met certain minimum standards.

- Make sure wires are securely fastened and that bare wire cannot touch anything except the connections it is meant to touch.

- Use the right materials for the job. Atmospheric conditions in barns (dust and humidity, for example) often make electrical appliances and fixtures that are designed for household use unsuitable.

—Wire should have *underground feeder* (UF) in its type designation if not run in a conduit and *wet area* (W) in its type designation if run in a conduit.

—Outlets should be protected with ground fault circuit interrupters (GFCIs or GFIs), and all lights, switches, and receptacles should be in dust-tight, watertight, corrosion-resistant enclosures. Regular fluorescent lights do not perform well in barns, as high humidity makes them difficult to start. Fluorescent bulbs with a silicon coating work better.

—Connectors (like wire straps) should be weatherproof and corrosion resistant.

- Surface mount all wiring rather than burying it in walls. This allows easy inspection and repair and minimizes rodent damage (which is a big reason barn electrical systems fail). Provide protection (conduit or wood overlay) for all wire within 10 feet of floor height. Conduit comes in metal or PVC plastic. Both are reasonably easy to work with, though PVC is easier. Cables (the word *cable* refers to two or more wires in the same sheathing) that are mounted without conduit should be attached to surfaces with plastic-coated staples or straps placed every 4½ feet and within 8 inches of junction/outlet boxes.

- Always triple-check your work.

> **CAUTION**
>
> When working on electrical systems, turn off the power at the main service panel and use a voltage tester to confirm that the power is indeed off before beginning work.

TYPES OF DUST- AND MOISTURE-PROOF FIXTURES

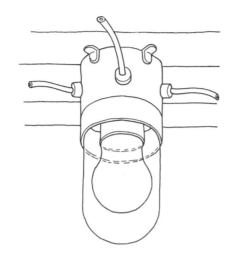

TIP

Mount light switch boxes on the latch side of doors for easy access.

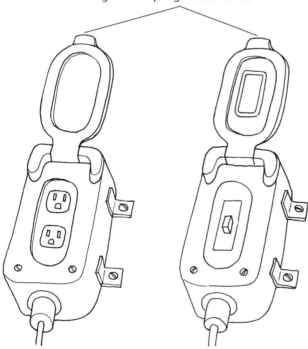

gasketed spring-loaded covers

Dust- and watertight boxes
with appropriate switch covers

Use only dust- and moisture-proof fixtures, like these, in your barn. The protective covers significantly reduce electrical system failures in the harsh barn environment, minimizing the chance of fire or electrocution.

Understanding Electricity

All matter is composed of combinations of atoms, and all atoms contain protons, neutrons, and electrons. An electrical charge is created when an electron moves from one atom to the next. Electrons associated with some atoms, such as copper, move easily, and so copper is known as a *conductor*. Most metals conduct electrons well; unfortunately, so do human bodies, making any high-voltage electrical work potentially deadly. Glass, plastic, and rubber don't conduct well, so they're considered *insulators*.

Electricity flows in a conductive circuit much the way blood flows through the body. For the flow to occur, there has to be a complete circuit. The hot wires are comparable to the arteries, taking power out to the extremi-

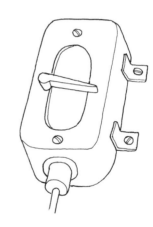

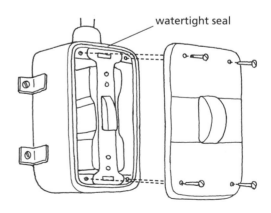

watertight seal

ties (the lights and appliances), and the neutral wires are like the blood vessels, returning power back to the heart (or current back to the main breaker box or battery).

Amperes, or amps (A), represent the quantity of current that can flow through a wire and is based on the size of the wire, but just as blood in an artery needs the pressure of a pump (the heart) to move it, so electricity needs a force to move it through the wire; this force is the *voltage*, or volts (V). A generator at the power plant, which acts as an electron pump, supplies voltage. The load (or power) of the system is measured in *wattage*, or watts (W). The watts are equal to the amps multiplied by the volts:

$$W = A \times V$$

and they are normally reported in units of watts called a kilowatt (kW), which is equal to 1000 watts. As electricity flows through the wire, there are losses due to *resistance*. The heat given off by lights and motors is actually the product of resistance.

Current delivered by the power company is three-phase, alternating current (AC), as opposed to direct current (DC), which is the type of current supplied by batteries. Utilities use alternating current at high voltages (for example, on the large metal towers that carry power from power plants to transformer substations, voltages can run in the hundred-thousand-plus range) because it has the ability to overcome resistance in the line. Each phase of current alternates in a sine wave pattern.

The current supplied through your meter has been "stepped down" from the voltage the electrical utility runs on its lines by means of a transformer attached to the pole, where the "service drop" comes off to enter the meter. (Transformers can also be used in some applications to step up voltage.) The vast majority of farms and homes are served with a three-wire, "single-phase" system that operates 120V and 240V appliances, fixtures, and motors, though some farms that are running larger loads have a separate four-wire, "three-phase" system that can operate a wider range of voltages found in commercial and industrial applications. On a three-wire system, two wires are hot — each rated at 120V — and the third is a neutral wire; 240V service is supplied by pulling from both hot wires. Lights and small appliances run off 120V, whereas larger appliances, like electric dryers, electric stoves, welders, and heaters, generally run off 240V.

From the meter, the electricity enters the main distribution panel (also known as the main breaker box or the main service entrance panel). Breakers are designed as a safeguard; they "trip" if they become overloaded, thereby stopping the flow of electricity. (Some older systems utilize fuses instead of breakers, but the fuse is designed to do the same thing.) From the main breaker, power is distributed to additional circuits via a series of circuit breakers and occasionally to additional breaker boxes, also called subpanels. Subpanels are required when all the breakers in the main panel are in use, or when a building is fed off a main service that is located in another building.

Breakers must be matched to wire size, and both breakers and wires matched to the load

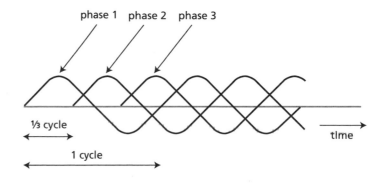

Power coming from the electrical utility is delivered as "three-phase" alternating current (AC). The currents vary over time in a sine wave pattern.

(the combined wattage of appliances and fixtures) that the "leg," or circuit, is expected to carry. Breakers are readily available in sizes from 15A up to 200A and come for 120V or 240V service. Most ordinary outlets and light circuits can be run on a 15A or 20A breaker; larger motors should have dedicated breakers that are sized at 125 percent of the amperage listed on the motor's nameplate. The larger the diameter of the wire, the less resistance there is, and therefore the greater the load it can carry. (Remember, resistance produces heat; therefore, an undersized wire carrying too great a load heats up and can ignite a fire.)

Wire size is described by a gauge number (the acronym AWG, American Wire Gauge, is used with the numbers), but the numbers are assigned in a counterintuitive fashion: the smaller the gauge number, the bigger the wire. Although this isn't usually an issue when running wires from the main panel to appliances and fixtures within a building, extremely long runs (say, from a main panel in the house to a new subpanel set in the barn) will need to have the wire up-sized by at least one gauge size to compensate for line losses from resistance.

The number and type of fixtures or appliances that can be served by a circuit breaker are figured from breaker size and voltage. For example, on a 20A, 120V circuit, a maximum of 2400W could be carried, but you should always leave at least a 30 percent reserve on each circuit for safety, so the practical load would be limited to 1680W. If this circuit is going to run a series of 300W heat lamps for brooding chickens, you could place up to five fixtures on it. If it will run regular lights and/or electrical outlets, figure no more than ten fixtures on a 20A circuit or seven on a 15A circuit.

Grounding

So, let's go back to the idea that electrical current is simply a series of electrons moving from atom to atom. The electrons move

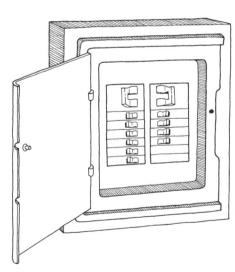

A breaker box is a crucial feature of any electrical system. It is a safety feature that cuts the flow of electricity to a circuit if a short develops, preventing too much current from flowing through the wires.

Electrons want to return to earth, and the grounding system is designed to provide them safe passage. When you (or your animals) touch an electrical fence, you act as a ground for electrons.

through the power grid because of the pressure placed behind them at the generator, but those "excited" electrons would rather just stay put with one atom and not keep moving. In other words, they would like to return to a neutral state. The best way for them to get back to a neutral state is to escape from the pressurized system of the wires and return to the earth, and they seek the easiest path to do so. The problem is that escaping electrical charges are dangerous if you happen to get in their way, as they are trying to get to "ground." (When you or your animals touch an electric fence and get a shock, you are providing the electrons with just such an easy escape route to the ground, but in the case of an electric fence, the charge is controlled in size and duration by the fence charger, so it's uncomfortable but not dangerous.)

Grounding systems are designed primarily as a safety feature; they facilitate escaping electrons (shorts) going back to the neutral state by providing a fairly safe route back to earth through the ground wires and electrode system. Electric utilities place ground lines frequently along power distribution systems (look at a series of power poles, and you will often see a heavy wire cable coming down the side of the pole to the ground) and at each transformer and meter. The system within your home, barn, or other outbuildings also has to be grounded, with each fixture and appliance (for example, motor housings) connected to a ground wire that returns to the "ground bus bar" in the breaker panel; the ground bus bar, in turn, connects the individual ground lines of circuits directly to the main ground line of the breaker panel. The

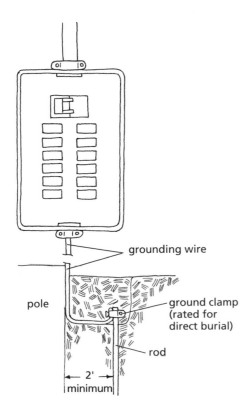

Proper grounding of fixtures, breaker boxes, and other electrical devices is crucial to their safe and proper operation. Grounding rods are driven into the earth, and grounding wires are attached to them.

TYPICAL ELECTRICAL LOADS	
Appliance/fixture	Watts
Air conditioner (room)	800–2500+*
Dryer (electric)	5000–12,000+*
Dryer (gas)	500*
Washer	600–2500*
Microwave	400–1200
Baseboard electric heat	250 per foot*
Space heater	600–5000+ (average 1500)
Fan	50–1500+*
Water heater (electric)	4500–15,000*
Motor	600–2400 per horsepower*
Lightbulb	40–150
Heat lamp bulb	300
Welder	10,000–50,000+*
Water pump	1000–2500*
Circular saw	2000*
Air compressor	4500*
Milking machine	12,600*
Bulk tank for cooking milk	31,500*
Silo unloader	16,500*
Gutter cleaner	6000*

*Uses significantly greater wattage when starting than when normally running; figures given include starting factors.

neutral line that comes into the main breaker box is continued through your internal system also, acting as the return line for the circuit (like the blood vessels returning blood flow to the heart), and it too is tied to the ground system at the main service breaker box so that there is an alternative way for electricity to go to earth during a failure in the system.

Faults in neutral and/or ground wiring are often responsible for fire in the worst case, or bizarre behavior of the electrical system in less severe cases. Lightbulbs that burn out frequently or dim when a motor kicks on and off are usually a sign of a neutral fault, or a neutral/ground fault, and are cause to immediately hire a licensed electrician to come in and troubleshoot the problem. Neutral faults — particularly in animal housing — are often the result of excessive dust coating the neutral wire or corrosion on it, which is one reason to use materials specifically designed for wet, dusty, and corrosive environments.

Although some people simply use a water line for grounding, it is best to run the main ground wire to several ground rods, driven at least 8 feet into the earth, with at least 16 feet between the rods. Locate them about 2 feet from the wall. In dry areas, drive longer rods (12 feet) and up to five rods to improve their effectiveness. Good grounding is critical to both safety (yours and your animals') and the effectiveness and longevity of all your electrical equipment, so do it right. To protect ground rods from damage, bury them so their tops are under about 6 inches of fill, but mark on your plans their exact location so you can find them easily in the future.

Laying It Out

Before you begin wiring, draw up a plan showing where switches, light fixtures, outlets, and motors for large equipment need to be, and also wire routes, points where junctions are needed, and wire size. The plan will help when you go shopping for materials. Codes control the location and minimum number of switches, light fixtures, and outlets for most buildings, but these items are inexpensive to install initially, so consider placing more than what the code calls for.

Providing adequate lighting is essential if you will routinely house animals in the barn. Plan on having at least one fixture per 150 square feet of pen area or one light per individual stall. Poultry housing should have either a dimmer system or a dual light system that allows for both low light and bright light

---

Stray Voltage

Ask any dairy farmer about stray voltage, and he or she will, at the very least, know that it is a problem in the dairy industry. Cows are quite sensitive to stray voltage, and their milk production is significantly reduced by regular jolts of it. Although other animals are less noticeably affected by stray voltage, they too can be negatively impacted.

Stray voltage occurs when there is a small voltage difference between a conductive surface and the ground. When an animal touches the conductive surface, like a metal water cup or stanchion, it thus provides a path for the current to move through its body to the ground. In other words, some of those excess electrons that are trying to find their way back to the earth make an intermediary trip through an animal's body. Animals that are being affected by stray voltage may avoid feeders or waterers, exhibit nervousness, or show reduced production (lower milk flow, slower growth, lessened fertility and reproduction).

Most stray voltage problems relate directly to neutral faults, ground faults, neutral-to-ground faults, or improperly grounded equipment. (Lightning can also cause stray voltage, but it is of short duration, so it is not the same type of persistent problem for livestock.) Unfortunately, one of the worst things about stray voltage is that the fault doesn't even have to be on your farm; it can be coming to your animals from a fault some distance away on another farm or from the utility company's distribution system.

Proper installation of all wiring and equipment, especially providing adequate grounding for all metal equipment in the barn, is the first step in preventing the problem. But if your animals are showing some of the symptoms suggesting stray voltage is a problem, hire an electrician who specializes in stray voltage diagnosis to ascertain if indeed it is a problem on your farm and help find the source. (In some areas, the power utility may have a staff person trained to diagnose stray voltage problems and propose corrections.) If the problem is not easily fixed, there are "suppression" devices commercially available that cancel out stray voltage.

(after the birds have settled down to sleep at night, use the low lights if you need to enter the coop; during winter, bright lights can be used to extend laying). Include floodlight fixtures on barn sides at strategic locations. Milk parlors and milk houses need ample lighting; figure on one fixture per 100 square feet. In open housing areas, make sure lights are located over feed bunks and water areas.

Installation

To install your system, work on one circuit at a time. First install all the switch, outlet, and junction boxes along the circuit. To the extent possible, place boxes in areas where animals can't access them, with switches mounted 4 feet off the floor and outlet boxes about 14 inches above the floor. When you need to place a box in an area that animals may have even occasional access to, the boxes should be protected and high enough to reduce the likelihood of animal injury. Junction boxes are used at points where wires are going to split in two different directions and where light fixtures are going to be set.

Each large, permanent motor should have its own disconnect box with a switch that's rated greater than the horsepower of the motor that will be attached to it and that has a current rating greater than 1.15 times the regular operating current. Most motors are expensive, so invest in switches that provide overload protection; they will shut off the motor before it overheats, preventing permanent and serious damage to the motor.

After the boxes are in place, start running wire from the breaker panel to the first box, but don't yet connect the wire to the circuit breaker or to switches, fixtures, or outlets. Leave about 20 inches of wire loose just outside the panel and about 8 inches of wire loose coming through the boxes for making connections so you have plenty of wire for making connections when the time comes. Strip about 6 inches of sheathing off the ends of the wires. Continue pulling wire to the rest of the boxes. (My approach is to tack the wire loosely in a few places while I pull it, and then, after it is out to the end of the line and I'm sure I have it laid out properly for all the

<aside>
CAUTION

Before working on a panel, make sure the main breaker is thrown to the OFF position, but remember that the wires coming into that breaker are still hot, so you must completely avoid any contact with them.
</aside>

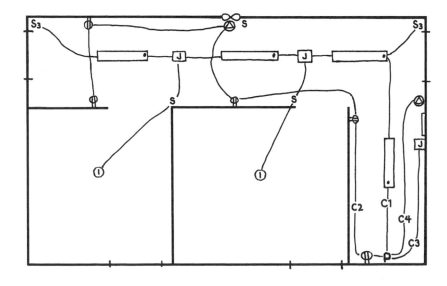

Draw up a plan to aid you in wiring. Show fixtures, junction boxes, switches, and outlets. Make sure circuits are not overloaded.

KEY

- ⌀ Duplex outlet
- ⌂ Special purpose outlet
- **S** Switch
- **S₃** Three-way switch
- **P** Breaker box/Service panel
- **J** Junction box
- ⓘ Incandescent fixture
- ▭ Fluorescent fixture

C1 Circuit 1
    4 Fluorescent
    2 Incandescent
C2 Circuit 2
    5 Duplex outlet
    1 Special purpose outlet
C3 Circuit 3
    To haymow lights
C4 Circuit 4
    1 220 outlet

TYPES OF ELECTRICAL CIRCUITS

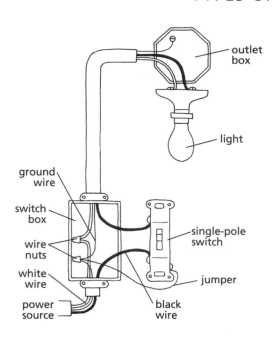

Basic wiring for a single switch-controlled light; other fixtures like outlets and fans can be controlled this way. The white wire is neutral; the black wire is hot.

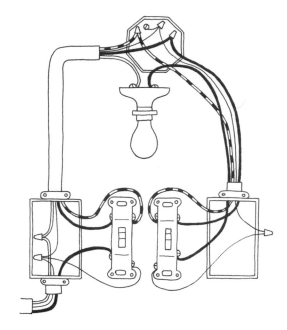

Sample wiring scheme for a light fixture controlled by two three-way switches; these switches are used to turn things on and off from multiple locations.

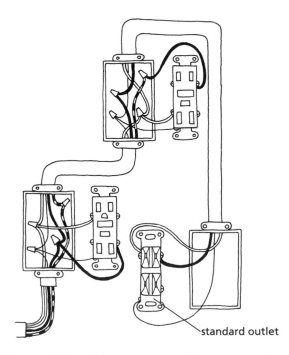

Good approach for wiring two GFCI outlets. GFCIs can be wired in series, but when they are, a fault on one outlet will trip all the GFCIs, making troubleshooting a challenge. Standard outlets, like the one shown in the lower right, can be pigtailed, as shown, into the circuit where needed.

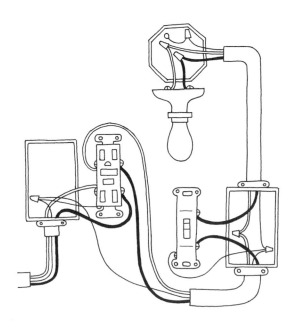

This diagram shows how to series wiring out of a GFCI outlet. This example shows how to wire a light and switch, but the approach is the same for any fixture in series after a GFCI outlet. When fixtures are run in series after a GFCI outlet, they will not work if the GFCI outlet trips.

connections, I go back and connect it firmly along its whole run with plastic-coated staples or straps.) Avoid crimping wire; it's possible to damage the wires within the sheathing without even knowing it.

Once the wire has been run, begin making connections from the far end of the circuit, working back toward the main panel. The easiest way to connect wires is by using "solderless wire connectors," or wire nuts. These are available in a variety of sizes that accommodate different sizes of wires and multi-wire installations. In most cases, you will be working on 120V circuits with a three-wire cable containing one hot wire (usually black or red), one neutral wire (white), and one bare (or sometimes green) copper wire for the ground. All metal boxes should be grounded. Switches and receptacles can be connected either via the screws that are attached to the sides or by pushing the wire into the self-clamping push holes in the back. For motors, you may need a 240V circuit. Motors are wired with two hots (one black and one red), one neutral, and one ground.

Once all the connections have been made along the circuit, you can make the connection to the main panel. In some municipalities or counties, this connection must be completed by a licensed electrician. Again, if you feel at all less than competent, hire a professional to check your work and make these connections.

Use a large screwdriver and a pair of pliers to remove the knockouts in the sides of the panel. Cable is connected to the panel using cable clamps. Connect the black wire to the circuit breaker, the white to the neutral strip, and the bare wire to the ground strip, starting at the top breakers and working down. Try to keep the wires neat inside the box so you can easily determine which wire goes where.

After the connections have been made, place the cover over the circuit breakers and accurately label the breakers. Labels are required by code, but even if they weren't, it's important to know which breaker controls which circuit in the barn.

Lightning Protection

Lightning strikes the earth eight million times per day. A typical bolt will discharge electrical current in the 1000A to 300,000A range, with up to 100 million volts behind it; the air temperature surrounding a bolt reaches 50,000°F, which is hotter than the sun's surface! It is a major cause of fire on farms and the cause of 80 percent of all livestock losses due to accidents. Buildings are particularly vulnerable if they are the only things around for some distance, as farm buildings often are.

A properly installed lightning protection system safely directs this massive current to the ground, providing almost 100 percent protection for structures. Most insurance companies reduce fire coverage rates on buildings containing these systems.

In such systems, lightning rods are placed at regular intervals on the roof and attached to a ground system by means of heavy cables. Arrestors prevent surges of electricity from entering the building wiring. Lightning protection systems should be installed by a contractor who has the experience and equipment to do so, using materials approved by the Underwriters Laboratories.

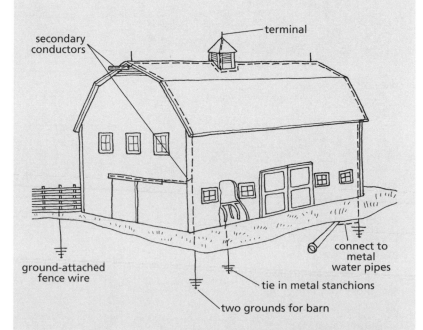

secondary conductors
terminal
ground-attached fence wire
connect to metal water pipes
tie in metal stanchions
two grounds for barn

Protecting property and animals from lightning with a lightning protection system is a good investment and will usually reduce your insurance premiums.

COLD-CLIMATE PLUMBING

Water systems must be designed and installed with protection in mind where winter temperatures potentially or regularly drop below freezing. Water and sewer lines must be buried below the frost line or insulated. Frost lines may reach as deep as 10 feet in the mountains and some northern states (and Canada). Lines need to be 2 feet deeper if they are under highly compacted areas like driveways or livestock holding pens than if they're under uncompacted, natural areas. Self-draining, sanitary, frost-free hydrants can be used; they're set in the ground to a depth below the frost line.

Within unheated buildings, lines must be designed to drain to a low point, with a valve. Water lines in buildings can be protected from freezing by using pipe insulation in combination with a heating cable or by using a device that runs the water intermittently depending on temperature.

Running a constant trickle during cold weather may protect the water line from freezing, but if the trickle discharges to a sewer line, it may freeze that line. In a building with a heated room, such as a milk room or office, the water line can be run into the heated area and a frost-proof hose bib can be run to the unheated area. These hose bibs are designed to drain out when the hose is disconnected so the ice doesn't move back into the heated part of the line.

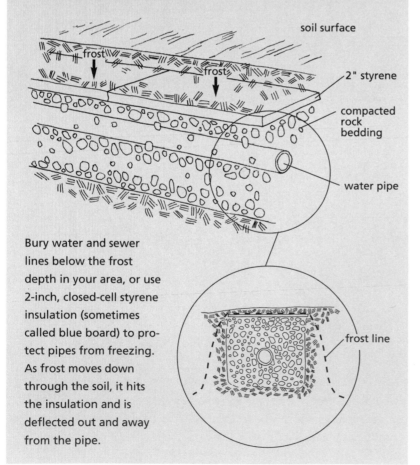

Bury water and sewer lines below the frost depth in your area, or use 2-inch, closed-cell styrene insulation (sometimes called blue board) to protect pipes from freezing. As frost moves down through the soil, it hits the insulation and is deflected out and away from the pipe.

Alarms

Consider installing an alarm (or alarms) in barns that will regularly house animals. Fire detection and electrical failure resulting in ventilation-system failure are probably the main concerns, though alarms can also be designed to go off in cases of toxic gas levels, equipment failure, or entrance by an intruder. Heat-detection alarms work better than conventional fire alarms in barns because dust and condensation can set off standard fire alarms with false positives. Some alarm systems have microprocessors and automatic telephone callers connected to emergency service providers (police/fire), whereas others just turn on a light or blow a siren.

PLUMBING

Although plumbing is a really wonderful thing and can be valuable in a barn, it adds complications to a building project. A system that provides a hose bib for filling stock tanks and buckets isn't too difficult, but for a system that will provide sinks, toilets, and other facilities, you will need both incoming water and outgoing wastewater systems, which add considerable expense to initial construction costs and may require the services of a licensed plumber, depending on your town regulations and confidence level.

There are two similarities between plumbing work and electrical work: First, all states require some inspection for plumbing, and many require the work be done by a licensed plumber. Also, as with the electrical supply, there's a lot to be said for the surface mounting of water-supply lines rather than burying them in walls. Accessible lines make repairing leaks quick and easy.

At points where underground water-supply pipes come to the surface in barns (or livestock pens), particularly where pipes penetrate concrete, cover them with foam expansion wrap. Then run them through a larger-diameter rigid pipe to protect them from damage by animals or frost-heaving concrete. This design also helps reduce freezing if the outer pipe goes a foot or so deeper than the frost line, because ground temperature can warm the air around the water-supply pipe.

The water supply to a barn should also have backflow prevention to protect drinking water from contamination. A backflow, or cross-connection, draws water back into the potable water supply when negative water pressure is applied, which can be caused by undersized or broken lines, high withdrawal rates, or elevation changes in the system. Vacuum breakers or double-check valves are useful to protect drinking water from bacterial or chemical contamination by preventing such backflows.

Water Supply

Most citizens in the United States are now supplied by municipal water systems, though many farms use a private well for water. If you receive city water, learn where your shutoff valve is and keep the cap to it accessible at all times, just in case. Consider acquiring a "key" to shut off the valve yourself in case of emergency; otherwise you'll be waiting for a city utility worker or plumber to show up and turn off the water. If you have a key, operate the valve once or twice a year to keep the mechanism operable.

City water providers are required to meet strict water-quality standards under the federal Safe Drinking Water Act, so you can usually count on consistently high-quality water if you are on a municipal supply. And, if at times the water is unsafe or not meeting standards, the city is required to notify users and/or supply an alternative source. Unfortunately, private wells don't have these same protections, and wells in rural areas often suffer contamination from pesticides, herbicides, fertilizers, or high bacterial levels. Consider having your well checked for contamination by a health department laboratory at least once every five years.

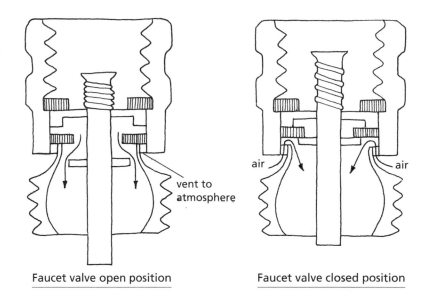

Faucet valve open position   Faucet valve closed position

Backflow preventers, like this atmospheric vacuum break, should be used on hose bibs, pumps, chemical mixers, and other devices that might allow contaminated water to flow back into your water piping. In the open position, water flows through the valve to the bib; In the closed position, air will enter the system and create a force that prevents backflow.

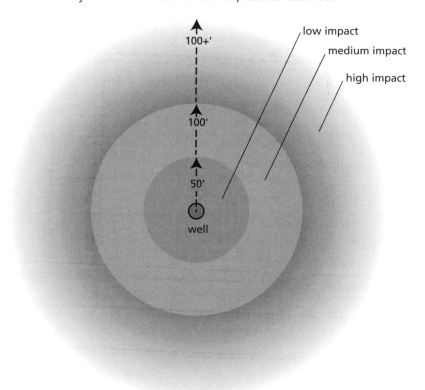

Wellhead protection is critical to your safety. The 50 feet nearest the well should be a low-impact area where no activities take place that might cause contamination. Between 50 and 100 feet, you could have a garage or house but not a feedlot or septic tank; these high-impact activities must be located at least 100 feet away from the well.

PIPES

Name	Acronym	Use	Fittings
Acrylonitrile butadene styrene	ABS	Sewer lines	Glue
Chlorinated polyvinyl chloride	CPVC	Good for corrosive water, and may be allowed for hot-water supply, depending on local codes	Glue
Polybutylene	PB	Suitable for hot- or cold-water supply	Heat, flare, or compression fittings
Polyethylene	PE	For cold-water supply or suitable for hot water to 100°F; often used in water heating in concrete floors; perforated PE is often used for drain lines around foundations	Specialized nylon or brass fittings, or stainless-steel clamps
Polyvinyl chloride	PVC	Cold water or sewer lines	Glue or specialty connections

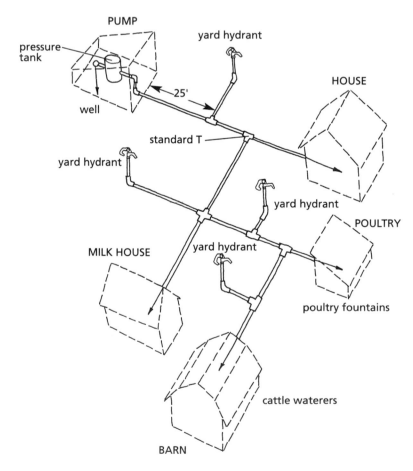

Water systems should be designed so that a separate line goes to each building, and each line should have its own shutoff valve. When installing lines, include plenty of yard hydrants in convenient locations.

The wellhead should be protected to minimize the chance of contamination. Most states have required setbacks (at least 100 feet) of septic systems from a wellhead, but always avoid any activities near a wellhead that can pollute your well. The well pipe should extend at least 8 inches above the ground, and soil should be banked to drain water away from the pipe.

City water is supplied with sufficient pressure that it does not require on-site storage, but private well systems generally require a pressure tank. The pressure tank serves two functions: it keeps the pump from running every time a small amount of water is needed, and it cushions the system from high-pressure surges created when the pump turns on. If the water supply is limited, private systems also may require an additional large storage tank, or cistern, that can be filled during times of low usage to provide adequate water during times of high usage. The pressure tank should supply at least 10 pounds per square inch (psi) of water pressure, though 20 psi is better.

Pipes

Whether you are installing a new service line from a city system or a well, you should install a shutoff valve for each branch line. By having individual shutoffs, or isolation valves, you can turn off water to one building for maintenance or repairs without shutting off water to all buildings.

Use a ¾-inch water-supply line to the barn. This provides good volume for filling stock tanks and cleaning purposes. Copper (rigid or flexible tubing), galvanized, or black-iron pipe tends to be relatively short-lived in barns due to corrosion. Plastic pipe with matched fittings withstands corrosion, and it can be used for cold-water installations; if your plumbing system will include a hot-water line, you may be required by code to use copper, although polybutylene is now making inroads for hot-water supply lines.

Copper and plastic are both relatively easy to work with and don't require highly specialized tools, though a wheel-style pipe cutter is a really good investment for do-it-yourself plumbing. Galvanized and black pipe have to be threaded, which requires more expensive, specialized tools. If you plan well before you go shopping, however, you can purchase prethreaded pipes in a variety of sizes for any job.

Be careful not to kink copper tubing or those points will be the first to fail. Pipes should be installed where they are not likely to be damaged by animals or sharp objects, and they should be secured so they don't sag. Use hangers (between 24 inches and 32 inches works well), or you may be able to run pipes through holes drilled in framing (which should be in at least 2 inches from the edge of the joists or studs, and as small as possible to allow the pipe to pass through but not more than one sixth the size of the board). If metal piping is used, the hangers should be of the same type of metal to reduce corrosion from galvanic action.

Copper pipe is connected to fittings by "sweating," or soldering. Lightly sand cut edges with an emery cloth to remove metal debris and provide a better fit, then coat the surfaces that are to be joined with soldering paste, or flux, and assemble them. Apply heat evenly over the surface with a handheld propane torch (wear gloves or use clamps to hold the heated metal). Once the copper is hot, touch the solder against the joint until it begins sucking the solder in. Flexible copper tubing should be joined with specialty compression fittings instead of solder.

Plastic pipe is used almost exclusively for drain, waste, and vent pipes (called DWV pipes), but it is still not common for water-supply lines, though Ken and I have had good luck with polybutylene for both hot- and cold-water supplies over the years and wouldn't hesitate to use it in a barn application if allowed to do so by the local plumbing code. It is easy to

CONNECTING COPPER PIPE

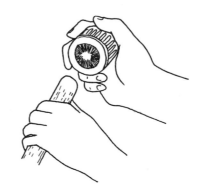

1. To prepare copper pipe for soldering, lightly sand the edges with an emery cloth or metal-tooth pipe sander (shown) to remove metal debris and provide a good surface for solder to adhere to.

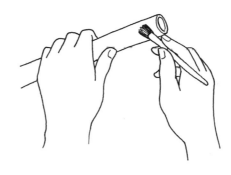

2. Apply a coating of soldering paste to the surface, then slide the fitting over the pipe.

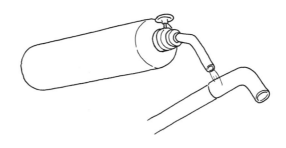

3. With a handheld propane torch, heat the metal until it is hot all the way around.

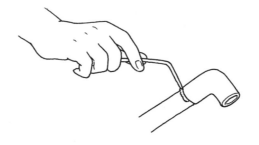

4. Apply the solder wire against the hot pipe. It should be drawn into the gap between the pipe and the fitting.

work with because it's joined to couplings by specialized nylon or brass connectors, and because of its flexibility, it is resistant to breaking when the water in it freezes.

PVC, which is the most common DWV pipe, is usually glued. The glue is actually a specialized solvent that welds the plastic together. For waste to flow properly out of drainage systems, the lines must be vented so that air pressure can keep things moving. If your property is serviced by a municipal wastewater-treatment plant, your barn will discharge to the sewer main in the street, but otherwise you will need to include a septic system. Septic systems have three primary components: the tank, the distribution box, and the leach field. All states require they be constructed to certain minimum standards and inspected. The system is sized based on the calculated loads. Plastic pipe that is buried as part of the waste-disposal system should be well bedded with clean gravel. To protect it

from breakage, bed below the pipe to a depth equal to the pipe's diameter and above the pipe to a depth equal to twice the pipe's diameter. The pipe should slope down by at least ¼ inch in 12 inches; use rigid styrene insulation to prevent freezing.

Two types of valves are commonly used in plumbing: gate valves and globe valves. Gate valves, which don't restrict flow, should be installed on incoming water-supply lines where they enter the building so the supply can be quickly shut off in case of emergency. They are also a good idea on the inlet side of appliances like water heaters, toilets, and washing machines. Globe valves are generally used on hose bibs and sinks, as they can control the flow of water from a trickle to full bore.

INTERIOR FINISH WORK: DRYWALL

Drywalling isn't difficult, but the sheets are heavy (a ½-inch sheet of 4x8 weighs over 50 pounds) and awkward to work with. If dropped, they break. Although it can be installed with nails, we prefer to use screws. With a battery-powered drill, screws are a snap to install, and they need to be placed only every 12 inches, whereas nails should be set every 8 inches. Sheets of drywall are 4 feet wide and come in a range of lengths, from 8 feet to 12 feet. Whenever possible, place sheets horizontally, and choose the longest length you can deal with, as it reduces the number of seams to be taped and plastered.

Taping and pasting drywall takes practice, but a barn is a good place to learn because a few imperfections won't be a serious problem. The tape is self-adhesive, and the plaster (also called drywall mud, joint compound, paste, or spackle) comes dry for mixing as needed or premixed in plastic tubs. I like the woven-fiber tape and the premixed paste because I find them a little bit easier to work with. Apply paste with a flat-edged putty knife to the tape and to nail or screw heads.

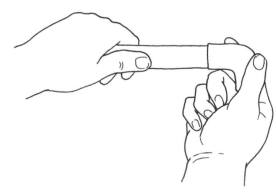

Gluing PVC is easy. First, brush pipe primer around the area on the pipe where the fitting will go, then apply glue (also called pipe cement) evenly around the same area. Push the fitting on and twist it one quarter to disperse glue across the fitting. Let dry for a few minutes.

INSTALLING DRYWALL

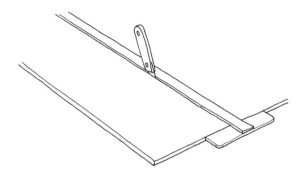

1. Dry wall is easy to cut with a utility knife. You can run a chalk line and follow it, or use a straight edge as a guide. Cut through one surface. Lift the board and bend it at the cut, then turn it over and cut through the paper backing at the seam.

2. It is best to install the top sheet first, then the bottom sheet if more than one sheet is required. A helper makes this job easier. If you don't have a helper, temporarily nail a 2x4 across the studs just below the bottom of the sheet to support it while you work.

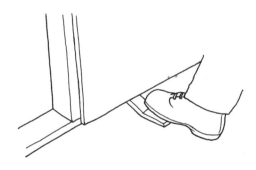

3. Push the bottom panel so it is flush against the top panel. Use a foot lifter for leverage to properly place the panel.

4. If you are using self-adhesive tape, just run it down the seams with your hand to keep it smooth. For plain paper tape, apply a thin coat of plaster to the seam, then run the tape along it, smoothing it with a putty knife.

5. Last, apply the final coat of putty. I like to use a wide putty knife for the final coat. Carefully smooth the putty and blend out the edges. When the putty is completely dry, sand it lightly to smooth the edges and make neat seams.

PLAN CREDITS

Most of the plans in this book were originally developed by the United States Department of Agriculture (USDA) and staff from the land-grant universities. The original plans may have included some additional information, like cutting lists or alternate views.

We have made some changes to the plans to reflect modern practices and materials, but the plans should also be reviewed by an agricultural engineer or building official, who can advise you about what modifications to the plans will be required to meet your needs and the current regulations in your region.

There are hundreds more of these older USDA/university plans. If you are interested in reviewing other plans, contact your Extension agent or see the resources for Web sites.

The credits that follow enumerate the original sources for the plans in this book.

Chapter 4
SMALL & PORTABLE HOUSING PROJECTS

pages 37–39, rabbit hutch: Redrawn from Cooperative Extension Work in Agriculture and Home Economics, State of Georgia, University of Georgia College of Agriculture and USDA, Hutches for Rabbits, plan 6137, 1972.

pages 40–41, two-compartment rounded-corner hutch: Redrawn from Cooperative Extension Work in Agriculture and Home Economics, State of Georgia, University of Georgia College of Agriculture and USDA, Hutches for Rabbits, plan 6137, 1972.

pages 42–43, rabbit house: Redrawn from Cooperative Extension Work in Agriculture and Home Economics, The Pennsylvania State University and USDA, Rabbit House, plan 6233, 1975.

page 44, removable hutches: Redrawn from Cooperative Extension Work in Agriculture and Home Economics, The Pennsylvania State University and USDA, Rabbit House, plan 6233, 1975.

pages 46–47, A-frame hut: Redrawn from Cooperative Extension Work in Home Economics and USDA, "A" Type Hog House, plan Ex. 5666, 1949.

pages 48–49, modified A-frame hut 1: Redrawn from Cooperative Extension Work in Home Economics and USDA, Modified "A" Type Hog House, plan Ex. 5272, n.d.

page 50, modified A-frame hut 2: Redrawn from Cooperative Extension Work in Agriculture and Home Economics, Department of Agricultural Engineering, University of Maryland and USDA, Hog House, Modified "A" Type, plan 6116, 1971.

pages 51–52, portable shelter 1: Redrawn from The Pennsylvania State University Agricultural Extension Service, Portable Shelter, plan 725-01, n.d.

pages 53–54, portable shelter 2: Redrawn from Cooperative Extension Work in Agriculture and Home Economics, Department of Agricultural Engineering, University of Maryland and USDA, Portable Farrowing Pen, plan Ex. 5770, 1954.

page 55, backyard pen 1: Redrawn from Cooperative Extension Work in Agriculture and Home Economics and USDA, Backyard Pigpen, plan 6305, 1978.

pages 56–57, calf hutch with movable paddock: Redrawn from The Pennsylvania State University Cooperative Extension Service, Calf Hutch with Movable Paddock, plan 723-214, 1976.

pages 59–60, poultry shed: Redrawn from Cooperative Extension Work in Agriculture and Home Economics, The Pennsylvania State University and USDA, Poultry House 10'x12', plan 6248, 1976.

pages 61–62, portable brooder house: Redrawn from Cooperative Extension Work in Agriculture and Home Economics and USDA, Portable Brooder House, plan Ex. 5730, 1951.

pages 64–65, summer range shelter: Redrawn from Cooperative Extension Work in Agriculture and Home Economics and USDA, Summer Range Poultry Shelter, plan Ex. 5400, n.d.

pages 66–68, portable stable: Redrawn from Cooperative Extension Work in Agriculture and Home Economics, Department of Agricultural Engineering, University of Maryland and USDA, Portable Stable for a Horse, plan 6082, 1969.

pages 69–71, movable shed (perspective of framing, floor plan, section A-A, guard rail detail, center post detail): Redrawn from Cooperative Extension Work in Agriculture and Home Economics, Department of Agricultural Engineering, University of Maryland and USDA, Movable Hog House, plan Ex. 5787, 1954; (section [through farrowing stall], section A-A [flap open]): redrawn from Cooperative Extension Work in Agriculture and Home Economics, Department of Agricultural Engineering, University of Maryland and USDA, Two-Pen Movable Hog House, plan Ex. 5821, 1956.

Chapter 5
WINDBREAKS & SHADE SHELTERS

pages 80–82, permanent shade structure: Redrawn from Cooperative Extension Work in Agriculture and Home Economics, Department of Agricultural Engineering, University of Maryland and USDA, 12'x16' Shade for Hogs, Portable or Permanent, plan Ex. 5816, 1955.

pages 83–84, combination windbreak & shade: Redrawn from *Beef Housing and Equipment Handbook*. Ames, IA: 1986.

pages 85–86, portable shade structure: Redrawn from USDA Natural Resources Conservation Service, plan AL-ENG-54, n.d.

pages 87–88, portable shade for hogs: Redrawn from Cooperative Extension Work in Agriculture and Home Economics, Department of Agricultural Engineering, University of Maryland and USDA, Portable Shade for Hogs, plan Ex. 5947, 1963.

Chapter 6
BARNS & STABLES

pages 90–93, gambrel barn (floor plan, side elevation, window section): Redrawn from Cooperative Extension Work in Agriculture and Home Economics and USDA, General Barn, plan Ex. 5554, n.d.; (partial framing, end elevation): redrawn from Cooperative Extension Work in Agriculture and Home Economics, Department of Agricultural Engineering, University of Maryland and USDA, Horse Barn, 10' Stall, plan Ex. 6170, 1973; (exterior elevation, Dutch door; door frame elevation; exterior elevation, double sliding door): redrawn from Cooperative Extension Work in Agriculture and Home Economics, State of North Dakota, North Dakota Agricultural College and USDA, Barn Door Details, plan Ex. 5631, 1947.

pages 94–95, gambrel-style small barn (floor plan): Redrawn from Cooperative Extension Work in Agriculture and Home Economics and USDA, Horse Barn, Gambrel Roof, plan 6262, 1976; (stall at wall with wood block floor, section, plan): redrawn from Cooperative Extension Work in Agriculture and Home Economics, The Pennsylvania State University and USDA, Horse Stalls, plan Ex. 5175, n.d.

pages 96–98, small gable barn: Redrawn from Cooperative Extension Work in Agriculture and Home Economics and USDA, General Barn, plan 6267, n.d.

pages 99–101, medium gable barn: Redrawn from Cooperative Extension Work in Agriculture and Home Economics, State of Tennessee, University of Tennessee Agricultural Engineering Department and USDA, General Barn, plan 6268, 1977.

pages 102–105, monitor barn: Redrawn from Cooperative Extension Work in Agriculture and Home Economics, State of Tennessee, The University of Tennessee Biosystems Engineering Department and USDA, 36'x48' General Barn (12' Center and 12' Shed, plan T4161, 1996.

pages 106–109, small stable: Redrawn from Agricultural Engineering Department, Cooperative Extension Service, University of Maryland, College Park — Eastern Shore, and USDA, Riding Horse Barn, plan Ex. 5838, 1957.

pages 110–112, larger stable: Redrawn from Cooperative Extension Work in Agriculture and Home Economics, The Pennsylvania State University and USDA, Horse Barn — 8 Stall, plan Ex. 6010, 1967.

pages 113–115, chicken coop: Redrawn from Cooperative Extension Work in Agriculture and Home Economics and USDA, Poultry House, plan Ex. 5580, n.d.

pages 117–119, walls, doors, and windows for pole-type buildings: Redrawn from Cooperative Extension Work in Agriculture and Home Economics, Department of Agricultural Engineering, University of Maryland and USDA, Walls, Doors, and Windows for Pole-Type Buildings, plan Ex. 5833, 1957.

pages 120–121 (top), basic loafing shed: Redrawn from Midwest Plan Service, Cooperative Extension Work in Agriculture and Home Economics and Agricultural Experiment Stations of North Central Region and USDA, 24' Frame Utility Building, plan 72044, 1961.

pages 121 (bottom)–123, sheep or goat shed: Redrawn from Cooperative Extension Work in Agriculture and Home Economics, Department of Agricultural Engineering, University of Maryland and USDA, Sheep Shed, plan Ex. 5811, 1955.

pages 124–126, enclosed shed: Redrawn from Cooperative Extension Work in Agriculture and Home Economics, The Pennsylvania State University and USDA, Sheep and Lambing Shed, plan Ex. 5919, 1961.

pages 127–129, sheep shed (floor plan): Redrawn from Cooperative Extension Service, Agriculture and Home Economics, The Pennsylvania State University and USDA, Sheep and Lambing Shed, plan 5919, 1961; (front elevation, end elevation, section A-A): redrawn from Cooperative Extension Work in Agriculture and Home Economics, Department of Agricultural Engineering, University of Maryland and USDA, Sheep Shed, Pole Construction, plan Ex. 5812, 1958.

pages 130–132, calf shed: Redrawn from Cooperative Extension Service, Agriculture and Home Economics and USDA, Solar Calf Shelter, plan 6374, 1985.

Chapter 7
ODDS & ENDS

page 137 (top right), poultry nest box: Redrawn from Cooperative Extension Work in Agriculture and Home Economics and USDA, Poultry Nests, plan Ex. 5077, 1933.

pages 143–144, lamb/kid brooder: Redrawn from Cooperative Extension Work in Agriculture and Home Economics, Department of Agricultural Engineering, University of Maryland and USDA, Combination Lamb Brooder and Ewe Feeder, plan Ex. 5863, 1958.

pages 146–147, milking barn: Redrawn from Cooperative Extension Work in Agriculture and Home Economics and USDA, Milking Barn and Milkhouse for 10 Goats, plan 6255, 1977.

pages 148–150, metal stanchion: Redrawn from Cooperative Extension Service, Agriculture and Home Economics and USDA, Goat Milking Stand — Metal, plan 6399, 1988.

pages 153–154, pig brooder: Redrawn from Cooperative Extension Work in Agriculture and Home Economics, Department of Agricultural Engineering, University of Maryland and USDA, Electric Brooder for Pigs, Hover Type for Incandescent Lamps, plan Ex. 5907, 1960.

pages 156 (bottom)–157, mineral feeder: Redrawn from Cooperative Extension Work in Agriculture and Home Economics, Department of Agricultural Engineering, University of Maryland and USDA, Salt and Mineral Box, plan Ex. 5769, 1952.

page 161 (middle), door partition wall detail: Redrawn from Cooperative Extension Work in Agriculture and Home Economics and USDA, Horse Barns for Hot Climate, plan 6337, 1981.

page 161 (bottom), stall partition, cutaway section: Redrawn from Cooperative Extension Work in Agriculture and Home Economics, The Pennsylvania State University and USDA, Riding Horse Barn, plan Ex. 5838, 1957.

RESOURCES

GOVERNMENT PUBLICATIONS

The MidWest Plan Service (MWPS) and Natural Resource, Agriculture, and Engineering Service (NRAES) are organizations of Extension and research agricultural engineers from universities in the Midwest and Northeast, with representatives of the USDA. They sell dozens of excellent publications that we have become big fans of over the years. They are available from county Extension agents or at *www.mwps.org*. Titles I recommend are:

NRAES-1: *Post-Frame Building Handbook*. Ithaca, NY, 1997.
NRAES-27: *Lumber from Local Woodlots*. Ithaca, NY, 1988.
MWPS-3: *Sheep Housing and Equipment Handbook*. Ames, IA, 1994.
MWPS-6: *Beef Housing and Equipment Handbook*. Ames, IA, 1987.
MWPS-35: *Farm and Home Concrete Book*. Ames, IA, 1989.

The U.S. Department of Housing and Urban Development (HUD) has published a Design Guide for Frost-Protected Shallow Foundations that is available at www.huduser.org/publications/destech/desguide.html.

BOOKS

Burch, Monte. *Building Small Barns, Sheds & Shelters*. North Adams, MA: Storey Publishing, 1983.
——— . *How to Build Small Barns and Outbuildings*. North Adams, MA: Storey Publishing, 1992.
——— . *Monte Burch's Pole Building Projects*. North Adams, MA: Storey Publishing, 1993.
Damerow, Gail. *Fences for Pasture and Garden*. North Adams, MA: Storey Publishing, 1992.
Hill, Cherry. *Horsekeeping on a Small Acreage*, Second Edition. North Adams, MA: Storey Publishing, in press.
——— . *Stablekeeping*. North Adams, MA: Storey Publishing, 2000.
Ekarius, Carol. *Small-Scale Livestock Farming*. North Adams, MA: Storey Publishing, 1999.
Engler, Nick. *Renovating Barns, Sheds & Outbuildings*. North Adams, MA: Storey Publishing, 2001.
Klimesh, Richard, and Cherry Hill. *Horse Housing: How to Plan, Build, and Remodel Barns and Sheds*. North Pomfret, VT: Trafalgar Square, 2002.
Sobon, Jack, and Roger Schroeder. *Timber Frame Construction*. North Adams, MA: Storey Publishing, 1984.

INTERNET RESOURCES

American Livestock Breeds Conservancy
www.albc-usa.org
Breed conservation

Barn Again!
National Trust for Historic Preservation
www.preservationnation.org/information-center/saving-a-place/rural-heritage/barn-again
How to rehab historic barns

Carol Ekarius
www.carolekarius.com
The author's personal website

Earth911.com
http://earth911.org
Recycling and disposal

Horsekeeping LLC
www.horsekeeping.com
Cherry Hill's website where she offers tips and advice

National Institute of Food and Agriculture
(formerly the Cooperative State Research, Education, and Extension Service)
United States Department of Agriculture
www.csrees.usda.gov
Search for Extension specialists and programs

TerraServer
www.terraserver.com
Topographical maps and aerial photos

TheCityChicken.com
www.thecitychicken.com
Helpful information on chickens

ThomasNet
www.thomasnet.com
Search a database of manufacturers and suppliers by keyword

Tools of the Trade **Magazine**
www.toolsofthetrade.net
Review of tools by the pros

PLANS FROM LAND-GRANT UNIVERSITIES
The following websites, maintained by land-grant universities and the National Institute of Food and Agriculture, are helpful resources for finding animal housing plans in PDF.

Biosystems Engineering & Soil Science Department
University of Tennessee Institute of Agriculture
http://bioengr.ag.utk.edu

Building Plans
North Dakota State University Extension Service
www.ag.ndsu.edu/extension-aben/buildingplans

Colorado State University Extension
www.csuextstore.com

Penn State Extension
Agricultural and Biological Engineering Department
http://abe.psu.edu/extension

PLANS FOR BARNS AND STABLES
Note: An asterisk (*) before the name of a firm indicates that it specializes in custom barn designs; all others are plan services with set designs.

Apple Valley Horse Barn Plans
www.applevalleybarns.com
This website highlights plans from various architects and designers

Ashland Barns
Central Point, Oregon
ashlandbarnplans4u@yahoo.com
www.ashlandbarns.com

BarnPlans, Inc.
North Bend, Oregon
877-259-7028
www.barnplans.com

BGS Plan Company
Roseville, California
877-752-6247
www.bgsplanco.com

*Blackburn Architects
Washington, District of Columbia
202-337-1755
San Francisco, California
415-439-5203
http://blackburnarch.com

*CMW, Inc.
Lexington, Kentucky
800-494-6623
www.cmwequine.com

The COOL House Plans Company, Inc.
Beaufort, South Carolina
800-482-0464
www.coolhouseplans.com

*Gralla Equine Architects
GH2 Architects, LLC
Tulsa, Oklahoma
918-587-6158
Norman, Oklahoma
405-701-1515
www.grallaarchitects.com

*Harrison Banks
Newton, Massachusetts
617-236-1876
www.harrisonbanks.com

*Hayward Designs, Ltd.
LaFayette, Georgia
423-488-0460
www.haywarddesigns.com

Homestead Design, Inc.
Anacortes, Washington
360-230-1917
www.homesteaddesign.com

StableWise, LLC
Duvall, Washington
425-788-4676
www.stablewise.com

BUILDERS AND KIT MANUFACTURERS SPECIALIZING IN BARNS AND STABLES

A & B Lumber Co. LLC
Pembroke, New Hampshire
800-267-0506
www.abbarns.com

All Buildings
Boca Raton, Florida
800-300-2470
Goleta, California
800-704-5484
www.allbldg.com

AmeriStall
Sanger, Texas
888-234-2276
www.ameristall.net

Amish Timber Framers, Inc.
Doylestown, Ohio
800-392-8789
www.amishtimberframers.com

Barns by Gardner
Loveland, Colorado
970-670-0707
www.barnsbygardner.com

BCI Barn Builders
Fort Gibson, Oklahoma
800-766-5793
www.bcibarns.com

Buildings and Barns, Inc.
Tempe, Arizona
800-316-0318
http://buildingsandbarns.com

Castlebrook Barns
Fontana, California
800-522-2767
www.castlebrookbarns.com

Classic Post & Beam
Kenduskeag, Maine
800-872-2326
www.classicpostandbeam.com

Cleary Building Corp.
Verona, Wisconsin
800-373-5550
www.clearybuilding.com

Conestoga Buildings
C. B. Structures, Inc.
New Holland, Pennsylvania
800-544-9464
www.cbstructuresinc.com

Country Carpenters, Inc.
Hebron, Connecticut
860-228-2276
http://countrycarpenters.com
Post and beam

F. C. P. Inc.
Wildomar, California
800-807-2276
www.fcpbuildings.com

Great Northern Barns
Canaan, New Hampshire
603-523-7134
www.greatnorthernbarns.com

Handi-Klasp Horse & Livestock Equipment
Weldy Enterprises
Wakarusa, Indiana
800-628-4728
www.weldyenterprises.com/ handi-klasp

Heritage Building Systems
North Little Rock, Arkansas
800-643-5555
www.heritagebuildings.com

Kentucky Steel Buildings Panel & Supply
Winchester, Kentucky
859-745-0606
www.kstbuild.com

Lester Buildings Systems, LLC
Lester Prairie, Minnesota
800-826-4439
www.lesterbuildings.com

MDBarnmaster
Corning, California
800-343-2276
www.mdbarnmaster.com

Morton Buildings, Inc.
Morton, Illinois
800-447-7436
www.mortonbuildings.com

National Barn Company
Central Division
Fort Gibson, Oklahoma
800-582-2276
www.nationalbarn.com

Norseman Structures Inc.
(formerly Cover-All Building Systems)
Saskatoon, Saskatchewan
855-385-2782
www.norsemanstructures.com
Buildings using patented truss-arch design and DuraWeave fabric covers

Port-A-Stall
Erie, Colorado
303-678-7838
http://port-a-stall.com

Walter's Buildings
Jack Walters & Sons, Corp.
Allenton, Wisconsin
800-558-7800
www.polebuilders.com

Wick Buildings
Mazomanie, Wisconsin
855-438-9425
www.wickbuildings.com

SUPPLIES, EQUIPMENT, AND MATERIALS

Acorn Manufacturing Company, Inc.
Mansfield, Massachusetts
800-835-0121
www.acornmfg.com
Forged-iron hardware

AG-CO Products
St. Johns, Michigan
800-522-2426
www.ag-co.com
Cupolas and stall accessories

Annapolis Weathervanes
Annapolis, Maryland
888-899-8493
www.weathervaneandcupola.com
Cupolas and weathervanes

ANSUL
Tyco Fire Protection Products
Marinette, Wisconsin
715-735-7411
www.ansul.com
Fire detection and prevention

Apollo Safety
Fall River, Massachusetts
800-813-5408
www.apollosafety.com
Fire detection and prevention

Barn Door Hardware
El Segundo, California
866-815-8151
www.barndoorhardware.com
Hinges, rollers, and hardware for barn doors

BARNWARE
Industrial Metal Products of Aberdeen, Inc.
Aberdeen, North Carolina
888-684-6773
www.barnware.com
Hardware and doors for barns

Behlen Mfg. Co.
Columbus, Nebraska
800-553-5520
www.behlenmfg.com

Bouvet USA
San Francisco, California
415-864-0273
www.bouvet.com
Traditional hardware

Buildings and Barns, Inc.
Mesa, Arizona
800-316-0318
www.buildingsandbarns.com
Variety of barn accessories

Country Manufacturing, Inc.
Fredericktown, Ohio
800-335-1880
www.countrymfg.com
Stall systems, waterers, feeders, etc.

Cumberland General Store
Alpharetta, Georgia
800-334-4640
www.cumberlandgeneral.com
Interesting odds and ends for around
the farm, including hand tools,
books, pumps, etc.

Dayville Hay & Grain, Inc.
Snohomish, Washington
800-407-2624
www.dayvillesupply.com
Variety of barn supplies

FICS of Maryland, Inc.
Sykesville, Maryland
800-378-4639
www.stableandarena.com

Gamewell-FCI
Honeywell
Northford, Connecticut
203-484-7161
www.gamewell-fci.com
Fire detection and prevention

**Handi-Klasp Horse & Livestock
Equipment**
Weldy Enterprises
Wakarusa, Indiana
800-628-4728
*www.weldyenterprises.com/
handi-klasp*
Stalls and other equipment

**Monroe Extinguisher
Company, Inc.**
Rochester, New York
585-235-3310
www.monroeextinguisher.com
Fire detection and prevention

MWI Components, Inc.
Spencer, Iowa
800-360-6467
www.mwicomponents.com
Roof vents and cupolas

National Bridle Shop, Inc.
Lewisburg, Tennessee
800-251-3474
www.nationalbridle.com
Tack room supplies

National Horse Stalls
Buena Vista, Virginia
800-903-8908
www.nationalhorsestalls.com
Stalls and doors

Nelson Manufacturing Company
Cedar Rapids, Iowa
888-844-6606
www.nelsonmfg.com
Waterers

New Concept Louvers
Springfield, Utah
800-635-6448
https://newconceptlouvers.com

Newer Spreader
Sanford, Florida
866-626-8732
www.newerspreader.com

North West Rubber Mats Ltd.
Abbotsford, British Columbia
800-663-8724
www.northwestrubber.com
Manufacturer of Red Barn Stall Mats

Port-A-Stall
Erie, Colorado
303-678-7838
http://port-a-stall.com
Pens and corrals

Ritchie Industries
Conrad, Iowa
800-747-0222
www.ritchiefount.com

SimplexGrinnel LP
Boca Raton, Florida
561-988-3600
www.simplexgrinnell.com
Fire detection and prevention

Summit Flexible Products
Dayton, Ohio
800-635-2044
www.groupsummit.com
Stall mats

INDEX

Numbers in *italics* indicate illustrations; numbers in **boldface** indicate charts.

Other Storey Titles You Will Enjoy

The Backyard Homestead, edited by Carleen Madigan.
A complete guide to growing and raising the most local food available
anywhere — from one's own backyard.
368 pages. Paper. ISBN 978-1-60342-138-6.

Chicken Coops, by Judy Pangman.
A collection of hen hideaways to spark your imagination and inspire you to
begin building.
176 pages. Paper. ISBN 978-1-58017-627-9. Hardcover. ISBN 978-1-58017-631-6.

Compact Cabins, by Gerald Rowan.
Simple living in 1,000 square feet or less — includes 62 design interpretations
for every taste.
216 pages. Paper. ISBN 978-1-60342-462-2.

Fences for Pasture & Garden, by Gail Damerow.
Sound, up-to-date advice and instruction to make building fences a task
anyone can tackle with confidence.
160 pages. Paper. ISBN 978-0-88266-753-9.

The Home Water Supply, by Stu Campbell.
Answers to all your questions about water: how to find it, filter it, store it, and
conserve it.
240 pages. Paper. ISBN 978-0-88266-324-1.

How to Build Small Barns & Outbuildings, by Monte Burch.
Complete plans and instructions for more than 20 projects, including
an add-on garage, a home office, a roadside stand, equipment sheds,
and four types of barns.
288 pages. Paper. ISBN 978-0-88266-773-7.

A Landowner's Guide to Managing Your Woods,
by Ann Larkin Hansen, Mike Severson & Dennis L. Waterman.
How to maintain a small acreage for long-term health, biodiversity, and high-
quality timber production.
304 pages. Paper. ISBN 978-1-60342-800-2.

Low-Cost Pole Building Construction, by Ralph Wolfe.
A definitive how-to book to save you money, labor, and time in building a small
home, barn, or other structure.
192 pages. Paper. ISBN 978-0-88266-170-4.

PlyDesign, by Philip Schmidt.
Distinctive plywood projects for every room in the house.
320 pages. Paper. ISBN 978-1-60342-725-8.

Rabbit Housing, by Bob Bennett.
Designs for building efficient facilities that can shelter two to one hundred
backyard rabbits in safety and comfort.
144 pages. Paper. ISBN 978-60342-966-5.

Reinventing the Chicken Coop, by Matthew Wolpe and Kevin McElroy.
Complete plans for 14 stylish, sustainable, and fully functional coops for every
level of builder.
176 pages. Paper. ISBN 978-1-60342-980-1.

Step-by-Step Outdoor Stonework, by Mike Lawrence.
More than 20 easy-to-build projects, illustrated with drawings and color
photographs.
96 pages. Paper. ISBN 978-0-88266-891-8.

Storey's Guide to Raising Series.
Everything you need to know to keep your livestock and your profits healthy.
All new editions of *Beef Cattle, Sheep, Pigs, Dairy Goats, Meat Goats, Chickens, Ducks, Turkeys, Poultry, Rabbits, Raising Horses, Training Horses,* and *Llamas.* New additions to the series: *Miniature Livestock* and *Keeping Honey Bees.*
Paper and hardcover. Learn more about each title by visiting *www.storey.com.*

Timber Frame Construction, by Jack Sobon and Roger Shroeder.
Clear explanations of the basics of timber frame construction.
208 pages. Paper. ISBN 978-0-88266-365-4.

The Vegetable Gardener's Book of Building Projects.
Simple-to-make projects, including cold frames, compost bins, planters, raised
beds, outdoor furniture, and more.
152 pages. Paper. ISBN 978-1-60342-526-1.

Woodworking FAQ, by Spike Carlsen.
Practical answers to common woodworking questions, plus insider tips on how
to be successful in every project.
304 pages. Paper with partially concealed wire-o. ISBN 978-1-60342-729-6.

These and other books from Storey Publishing are available
wherever quality books are sold or by calling 1-800-441-5700.
Visit us at *www.storey.com* or sign up for our newsletter
at *www.storey.com/signup.*